Evolutionary Biology

Evolutionary Biology

Edited by
Solomon Stevens

Larsen & Keller
www.larsen-keller.com

Evolutionary Biology
Edited by Solomon Stevens
ISBN: 978-1-63549-116-6 (Hardback)

目 Larsen & Keller

Published by Larsen and Keller Education,
5 Penn Plaza,
19th Floor,
New York, NY 10001, USA

Cataloging-in-Publication Data

Evolutionary biology / edited by Solomon Stevens.
      p. cm.
Includes bibliographical references and index.
ISBN 978-1-63549-116-6
1. Evolution (Biology). 2. Biology. 3. Evolution. I. Stevens, Solomon.
QH366.2 .E96 2017
575--dc23

The publisher's policy is to use permanent paper from mills that operate a sustainable forestry policy. Furthermore, the publisher ensures that the text paper and cover boards used have met acceptable environmental accreditation standards.

Printed and bound in the United States of America.

For more information regarding Larsen and Keller Education and its products, please visit the publisher's website www.larsen-keller.com

Table of Contents

Preface

Evolutionary biology refers to the study of evolutionary processes that result in the diversity of living organisms on Earth. It includes the study of the origin of life, decline of species and origin of new species. It is a sub-field of biology and is based on the principles of ecology, genetics, paleontology, systematics, etc. This book is a valuable compilation of topics, ranging from the basic to the most complex theories and principles in the field of evolutionary biology. Through this text, we attempt to further enlighten the readers about the new concepts in this field. It aims to serve as a resource guide for students and facilitate the study of the discipline.

Given below is the chapter wise description of the book:

Chapter 1- This chapter will provide an integrated understanding of evolutionary biology. Evolutionary biology studies the evolutionary process of the Earth and the diversity of life that originated from a single instance. The content on evolutionary biology offers an insightful focus, keeping in mind the complex subject matter.

Chapter 2- Evolutionary biology is an interdisciplinary subject. The genetic difference within and between populations is population genetics while evolutionary biology studies the evolutionary process that has produced the diversity of life. This chapter is a compilation of population genetics and evolutionary ecology.

Chapter 3- The sharing of a recent common ancestor by a group of organisms is known as common descent and adaption is the trait that has been maintained by the organism by the means of natural selection. Speciation and modern evolution synthesis are some other significant key concepts of evolutionary biology. The chapter strategically encompasses and incorporates the major components and key concepts of evolutionary biology.

Chapter 4- Natural selection is the key mechanism of evolution while heredity is the genetic information that is passed on from the parents to the offspring. The principles of evolutionary biology are natural selection, heredity and genotype. This chapter helps in understanding the diverse topics related to evolutionary biology.

Chapter 5- The change in the characteristics of a biological population over a period of years is called evolution whereas the process of change in the composition of sequences such as DNA, RNS and proteins is known as molecular evolution. This chapter on evolution offers an insightful focus, keeping in mind the complex subject matter.

Chapter 6- Biology concerns itself with the study of life, evolution and distribution of a living organism. The etymological meaning of biology in Greek is life; the origins of modern biology can easily be traced back to the Greeks. The chapter serves as a source to understand the history on the evolution of biology, although modern biology is a relatively recent development, sciences included within it have been studied since ancient times.

At the end, I would like to thank all those who dedicated their time and efforts for the successful completion of this book. I also wish to convey my gratitude towards my friends and family who supported me at every step.

Editor

Introduction to Evolutionary Biology

This chapter will provide an integrated understanding of evolutionary biology. Evolutionary biology studies the evolutionary process of the Earth and the diversity of life that originated from a single instance. The content on evolutionary biology offers an insightful focus, keeping in mind the complex subject matter.

Evolutionary Biology

Evolutionary biology is the subfield of biology that studies the evolutionary processes that produced the diversity of life on Earth starting from a single origin of life. These processes include the descent of species, and the origin of new species.

The discipline emerged through what Julian Huxley called the synthesis of understanding from several previously unrelated fields of biological research, including genetics, ecology, systematics and paleontology.

Current research has widened to cover the genetic architecture of adaptation, molecular evolution, and the different forces that contribute to evolution including not only natural selection but sexual selection, genetic drift and biogeography. The newer field of evolutionary developmental biology ("evo-devo") investigates how organisms develop (from a single cell through an embryo to an adult body) to find out the ancestral relationships between organisms and how the processes of development evolved.

Subfields

The study of evolution is the unifying concept in evolutionary biology. Evolutionary biology is a conceptual subfield of biology that intersects with other subfields that are delimited by biological organisation level (e.g., cell biology, population biology), taxonomic level (e.g., zoology, ornithology, herpetology) or angle of approach (e.g., field biology, theoretical biology, experimental evolution, palaeontology). Usually, these intersections are combined into specific fields such as evolutionary ecology and evolutionary developmental biology.

History

Evolutionary biology, as an academic discipline in its own right, emerged during the period of the modern evolutionary synthesis in the 1930s and 1940s (Smocovitis, 1996). It was not until the 1970s and 1980s, however, that a significant number of universities had departments that specifically included the term *evolutionary biology* in their titles, often in conjunction with ecology and behaviour. In the United States, as a result of the rapid growth of molecular and cell biology, many universities have split (or aggregated) their biology departments into *molecular and cell biology*-style departments and *ecology and evolutionary biology*-style departments (which often have subsumed older departments in botany, zoology and the like). The subdiscipline of palaeontology is often found in Earth science/geology/geoscience departments.

The statistician Ronald Fisher (1890 – 1962) helped to form the modern evolutionary synthesis of Mendelian genetics and natural selection.

J. B. S. Haldane (1892 – 1964) helped to create the field of population genetics.

Microbiology has recently developed into an evolutionary discipline. It was originally ignored due to the paucity of morphological traits and the lack of a species concept in microbiology. Now, evolutionary researchers are taking advantage of a more extensive understanding of microbial physiology, the ease of microbial genomics, and the quick generation time of some microbes to answer evolutionary questions. Similar features have led to progress in viral evolution, particularly for bacteriophages.

Many biologists have contributed to our current understanding of evolution. Although the term had been used sporadically starting at the turn of the century, evolutionary biology in a disciplinary sense gained currency during the period of "the evolutionary synthesis" (Smocovitis, 1996). Theodosius Dobzhansky and E. B. Ford were important in the establishment of an empirical research programme for evolutionary biology as were theorists Ronald Fisher, Sewall Wright and J. S. Haldane. Ernst Mayr, George Gaylord Simpson and G. Ledyard Stebbins were also important discipline-builders during the modern synthesis, in the fields of systematics, palaeontology and botany, respectively. Through training many future evolutionary biologists, James Crow, Richard Lewontin, Dan Hartl, Marcus Feldman, and Brian Charlesworth have also made large contributions to building the discipline of evolutionary biology.

Current research in evolutionary biology covers diverse topics, as should be expected given the centrality of evolution to understanding biology. Modern evolutionary biology incorporates ideas from diverse areas of science, such as molecular genetics and even computer science.

First, some fields of evolutionary research try to explain phenomena that were poorly accounted for by the work of the modern evolutionary synthesis. These phenomena include speciation, the evolution of sexual reproduction, the evolution of cooperation, the evolution of ageing, and evolvability.

Second, biologists ask the most straightforward evolutionary question: "what happened and when?". This includes fields such as palaeobiology, as well as systematics and phylogenetics.

Third, the modern evolutionary synthesis was devised at a time when nobody understood the molecular basis of genes. Today, evolutionary biologists try to determine the genetic architecture of interesting evolutionary phenomena such as adaptation and speciation. They seek answers to questions such as how many genes are involved, how large are the effects of each gene, to what extent are the effects of different genes interdependent, what sort of function do the genes involved tend to have, and what sort of changes tend to happen to them (e.g., point mutations vs. gene duplication or even genome duplication). Evolutionary biologists try to reconcile the high heritability seen in twin studies with the difficulty in finding which genes are responsible for this heritability using genome-wide association studies.

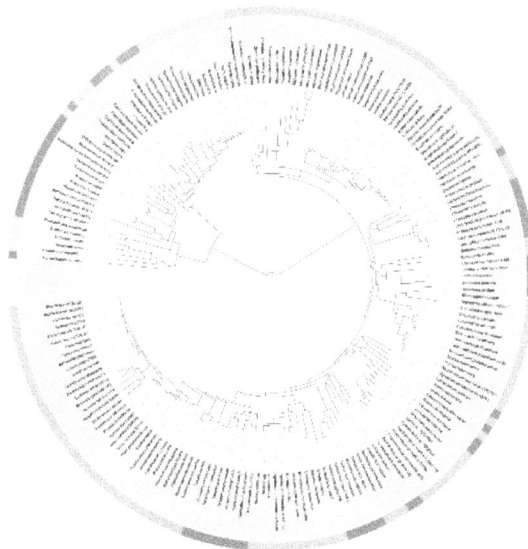

Graphical representation of the modern "Tree of Life on the Web" project.

One challenge in studying genetic architecture is that the classical population genetics that catalyzed the modern evolutionary synthesis must be updated to take into account modern molecular knowledge. This requires a great deal of mathematical development to relate DNA sequence data to evolutionary theory as part of a theory of molecular evolution. For example, biologists try to infer which genes have been under strong selection by detecting selective sweeps.

Fourth, the modern evolutionary synthesis involved agreement about which forces contribute to evolution, but not about their relative importance. Current research seeks to determine this. Evolutionary forces include natural selection, sexual selection, genetic drift, genetic draft, developmental constraints, mutation bias and biogeography.

An evolutionary approach is also key to much current research in biology that does not set out to study evolution per se, especially in organismal biology and ecology. For example, evolutionary thinking is key to life history theory. Annotation of genes and their function relies heavily on comparative, i.e., evolutionary, approaches. The field of evolutionary developmental biology ("evo-devo") investigates how developmental processes work by using the comparative method to determine how they evolved.

Journals

Some scientific journals specialise exclusively in evolutionary biology as a whole, including the journals *Evolution*, *Journal of Evolutionary Biology*, and *BMC Evolutionary Biology*. Some journals cover sub-specialties within evolutionary biology, such as the journals *Systematic Biology*, *Molecular Biology and Evolution* and its sister journal *Genome Biology and Evolution*, and *Cladistics*.

Other journals combine aspects of evolutionary biology with other related fields. For example, *Molecular Ecology*, *Proceedings of the Royal Society of London Series B*, *The American Naturalist* and *Theoretical Population Biology* have overlap with ecology and other aspects of organismal biology. Overlap with ecology is also prominent in the review journals *Trends in Ecology and Evolution* and *Annual Review of Ecology, Evolution, and Systematics*. The journals *Genetics* and *PLoS Genetics* overlap with molecular genetics questions that are not obviously evolutionary in nature.

Evolutionary Developmental Biology

Evolutionary developmental biology (evolution of development or informally, evo-devo) is a field of biology that compares the developmental processes of different organisms to determine the ancestral relationship between them, and to discover how developmental processes evolved. It addresses the origin and evolution of embryonic development; how modifications of development and developmental processes lead to the production of novel features, such as the evolution of feathers; the role of developmental plasticity in evolution; how ecology impacts development and evolutionary change; and the developmental basis of homoplasy and homology.

Although interest in the relationship between ontogeny and phylogeny extends back to the nineteenth century, the contemporary field of evo-devo has gained impetus from the discovery of genes regulating embryonic development in model organisms. General hypotheses remain hard to test because organisms differ so much in shape and form.

Nevertheless, it now appears that just as evolution tends to create new genes from parts of old genes (molecular economy), evo-devo demonstrates that evolution alters developmental processes to create new and novel structures from the old gene networks (such as bone structures of the jaw deviating to the ossicles of the middle ear) or will conserve (molecular economy) a similar program in a host of organisms such as eye development genes in molluscs, insects, and vertebrates. Initially the major interest has been in the evidence of homology in the cellular and molecular mechanisms that regulate body plan and organ development. However, subsequent approaches include developmental changes associated with speciation.

Basic Principles

Charles Darwin's theory of evolution builds on three principles: natural selection, heredity, and variation. At the time that Darwin wrote, the principles underlying heredity and variation were poorly understood. In the 1940s, however, biologists incorporated Gregor Mendel's principles of genetics to explain both, resulting in the modern synthesis. It was not until the 1980s and 1990s, however, when more comparative molecular sequence data between different kinds of organisms was amassed and detailed, that an understanding of the molecular basis of the developmental mechanisms began to form.

Currently, it is well understood how genetic mutation occurs. However, developmental mechanisms are not understood sufficiently to explain which kinds of phenotypic variation can arise in each generation from variation at the genetic level. Evolutionary developmental biology studies how the dynamics of development determine the phenotypic variation arising from genetic variation and how that affects phenotypic evolution (especially its direction). At the same time evolutionary developmental biology also studies how development itself evolves.

Thus the origins of evolutionary developmental biology come both from an improvement in molecular biology techniques as applied to development, and from the full appreciation of the limitations of classic neo-Darwinism as applied to phenotypic evolution. Some evo-devo researchers see themselves as extending and enhancing the modern synthesis by incorporating the findings of molecular genetics and developmental biology into an extended evolutionary synthesis.

Evolutionary developmental biology can be distinguished from earlier approaches to evolutionary theory by its focus on a few crucial ideas. One of these is modularity: as has been long recognized, plants and animal bodies are modular: they are organized into developmentally and anatomically distinct parts. Often these parts are repeated, such as fingers, ribs, and body segments. Evo-devo seeks the genetic and evolutionary basis for the division of the embryo into distinct modules, and for the partly independent development of such modules.

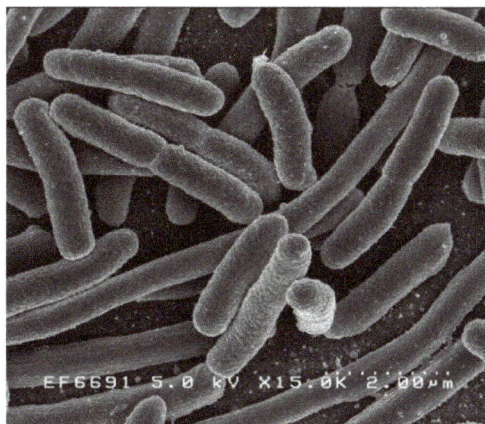

Escherichia coli

Another central idea recognizes that some gene products function as switches whereas others act as diffusible signals. Genes specify proteins, some of which act as structural components of cells and others as: hormones, receptors and enzymes that regulate various biochemical pathways within an organism. Most biologists working within the modern synthesis assumed that an organism is a straightforward reflection of its component genes. The modification of existing, or evolution of new, biochemical pathways (and, ultimately, the evolution of new species of organisms) depended on specific genetic mutations. In 1961, however, Jacques Monod, Jean-Pierre Changeux and François Jacob discovered within the bacterium Escherichia coli a gene that functioned only when "switched on" by an environmental stimulus. Later, scientists discovered specific genes in animals (including a subgroup of the genes which contain the homeobox DNA motif, called Hox genes) that act as switches for other genes, and could be induced by other gene products, morphogens, that act analogously to the external stimuli in bacteria. These discoveries drew biologists' attention to the fact that genes can be selectively turned on and off, rather than being always active, and that highly disparate organisms (for example, fruit flies and human beings) may use the same genes for embryogenesis, just regulating them differently.

Similarly, organismal form can be influenced by mutations in promoter regions of genes, those DNA sequences at which the products of some genes bind to and control the activity of the same or other genes, not only protein-specifying sequences. This finding suggested that the crucial distinction between different species (even different orders or phyla) may be due less to differences in their content of gene products than to differences in spatial and temporal *expression* of conserved genes. The implication that large evolutionary changes in body morphology are associated with changes in gene regulation, rather than with the evolution of new genes, suggested that Hox and other "switch" genes may play a major role in evolution, something that contradicts the neo-Darwinian synthesis.

Another focus of evo-devo is developmental plasticity, the basis of the recognition that organismal phenotypes are not uniquely determined by their genotypes. If generation of phenotypes is conditional, and dependent on external or environmental inputs, evolution can proceed by a "phenotype-first" route, with genetic change following, rather than initiating, the formation of changes in the phenotype such as to body structures (morphology). Mary Jane West-Eberhard argued the case for this in her 2003 book *Developmental plasticity and evolution*.

History

An early version of recapitulation theory, also called the *biogenetic law* or *embryological parallelism,* was put forward by Étienne Serres in 1824–26 as what became known as the "Meckel-Serres Law" which attempted to provide a link between comparative embryology and a "pattern of unification" in the organic world. It was supported by Étienne Geoffroy Saint-Hilaire as part of his ideas of idealism, and became a prominent part of his version of Lamarckism leading to disagreements with Georges Cuvier. It was widely supported in the Edinburgh and London schools of higher anatomy around 1830, notably by Robert Edmond Grant, but was opposed by Karl Ernst von Baer's embryology of divergence in which embryonic parallels only applied to early stages where the embryo took a general form, after which more specialised forms diverged from this shared unity in a branching pattern. The anatomist Richard Owen used this to support his idealist concept of species as showing the unrolling of a divine plan from an archetype, and in the 1830s attacked the transmutation of species proposed by Lamarck, Geoffroy and Grant. In the 1850s Owen began to support an evolutionary view that the history of life was the gradual unfolding of a teleological divine plan, in a continuous "ordained becoming", with new species appearing by natural birth.

In *On the Origin of Species* (1859), Charles Darwin proposed evolution through natural selection, a theory central to modern biology. Darwin recognised the importance of embryonic development in the understanding of evolution, and the way in which von Baer's branching pattern matched his own idea of descent with modification:

We can see why characters derived from the embryo should be of equal importance with those derived from the adult, for a natural classification of course includes all ages.

Ernst Haeckel (1866), in his endeavour to produce a synthesis of Darwin's theory with Lamarckism and Naturphilosophie, proposed that "ontogeny recapitulates phylogeny," that is, the development of the embryo of every species (ontogeny) fully repeats the evolutionary development of that species (phylogeny), in Geoffroy's linear model rather than Darwin's idea of branching evolution. Haeckel's concept explained, for example, why humans, and indeed all vertebrates, have gill slits and tails early in embryonic development. The recapitulation theory has since been discredited. However, it served as a backdrop for a renewed interest in the evolution of development after the modern evolutionary synthesis was established (roughly 1936 to 1947). Two of Haeckel's other ideas about the evolution of development have fared better: he argued in the 1870s that changes in the timing (heterochrony) and positioning within the body (heterotopy) of aspects of embryonic development would drive evolution by changing the shape of a descendant's body compared to an ancestor's. It took a century before evo-devo showed these ideas to be correct.

Stephen Jay Gould called the recapitulation explanation of evolution "terminal addition"; as if every evolutionary advance was added as a new stage by reducing the duration of all the older stages. The idea was based on observations of neoteny. Haeckel's heterochrony is a more general mechanism for evolutionary change, allowing any timings to change in any combination.

D'Arcy Thompson postulated that differential growth rates could produce variations in form in his 1917 book *On Growth and Form*. He showed the underlying similarities in *body plans* and how geometric *transformations* could be used to explain the variations.

Edward B. Lewis discovered homeotic genes, rooting the emerging discipline of evo-devo in molecular genetics. In 2000, a special section of the Proceedings of the National Academy of Sciences (PNAS) was devoted to "evo-devo", and an entire 2005 issue of the Journal of Experimental Zoology Part B: Molecular and Developmental Evolution was devoted to the key evo-devo topics of evolutionary innovation and morphological novelty.

The Developmental-Genetic Toolkit

The developmental-genetic toolkit consists of a small fraction of the genes in an organism's genome whose products control its development. These genes are highly conserved among phyla. Differences in deployment of toolkit genes affect the body plan and the number, identity, and pattern of body parts. The majority of toolkit genes are components of signaling pathways, and encode for the production of transcription factors, cell adhesion proteins, cell surface receptor proteins (and signalling ligands that bind to them), and secreted morphogens, all of these participate in defining the fate of undifferentiated cells, generating spatial and temporal patterns, which in turn form the body plan of the organism. Among the most important of the toolkit genes are those of the Hox gene cluster, or complex. Hox genes, transcription factors containing the more broadly distributed homeobox protein-binding DNA motif, function in patterning the body axis. Thus, by combinatorial specifying the identity of particular body regions, Hox genes determine where limbs and other body segments will grow in a developing embryo or larva. A paradigmatic toolkit gene is *Pax6/eyeless*, which controls eye formation in all animals. It has been found to produce eyes in mice and *Drosophila*, even if mouse *Pax6/eyeless* was expressed in *Drosophila*.

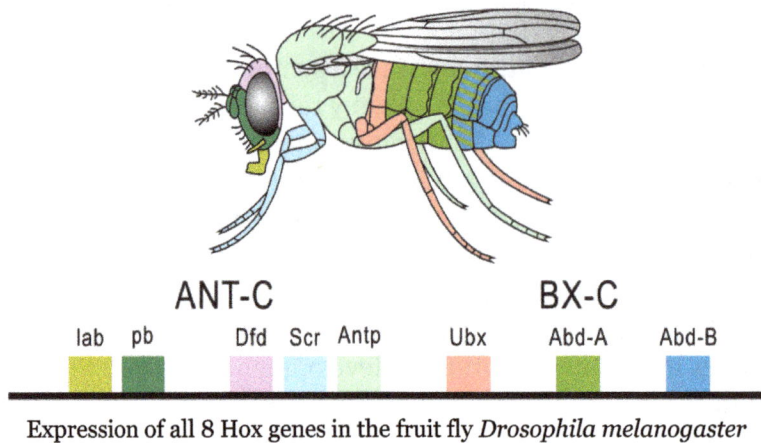

Expression of all 8 Hox genes in the fruit fly *Drosophila melanogaster*

This means that a big part of the morphological evolution undergone by organisms is a product of variation in the genetic toolkit, either by the genes changing their expression pattern or acquiring new functions. A good example of the first is the enlargement of the beak in Darwin's large ground-finch (*Geospiza magnirostris*), in which the gene *BMP* is responsible for the larger beak of this bird, relative to the other finches.

The loss of legs in snakes and other squamates is another good example of genes changing their expression pattern. In this case the gene *Distal-less* is very under-expressed, or not expressed at all, in the regions where limbs would form in other tetrapods. This same gene determines the eyespot pattern in butterfly wings, which shows that toolbox genes can change their function.

Toolbox genes, as well as being highly conserved, also tend to evolve the same function convergently or in parallel. Classic examples of this are the already mentioned *Distal-less* gene, which is responsible for appendage formation in both tetrapods and insects, or, at a finer scale, the generation of wing patterns in the butterflies *Heliconius erato* and *Heliconius melpomene*. These butterflies are Müllerian mimics whose coloration pattern arose in different evolutionary events, but is controlled by the same genes. This supports Marc Kirschner and John C. Gerhart's theory of Facilitated Variation, which states that morphological evolutionary novelty is generated by regulatory changes in various members of a large set of conserved mechanisms of development and physiology.

Development and the Origin of Novelty

Among the more surprising and, perhaps, counterintuitive (from a neo-Darwinian viewpoint) results of recent research in evolutionary developmental biology is that the diversity of body plans and morphology in organisms across many phyla are not necessarily reflected in diversity at the level of the sequences of genes, including those of the developmental genetic toolkit and other genes involved in development. Indeed, as Gerhart and Kirschner have noted, there is an apparent paradox: "where we most expect to find variation, we find conservation, a lack of change".

Even within a species, the occurrence of novel forms within a population does not generally correlate with levels of genetic variation sufficient to account for all morphological diversity. For example, there is significant variation in limb morphologies amongst salamanders and in differences in segment number in centipedes, even when the respective genetic variation is low.

A major question then, for evo-devo studies, is: If the morphological novelty we observe at the level of different clades is not always reflected in the genome, where does it come from? Apart from neo-Darwinian mechanisms such as mutation, translocation and duplication of genes, novelty may also arise by mutation-driven changes in gene regulation. The finding that much biodiversity is not due to differences in genes, but rather to alterations in gene regulation, has introduced an important new element into evolutionary theory. Diverse organisms may have highly conserved developmental genes, but highly divergent regulatory mechanisms for these genes. Changes in gene regulation are "second-order" effects of genes, resulting from the interaction and timing of activity of gene networks, as distinct from the functioning of the individual genes in the network.

A Genetic Theory of Morphological Evolution

The discovery of the homeotic Hox gene family in vertebrates in the 1980s allowed researchers in developmental biology to empirically assess the relative roles of gene duplication and gene regulation with respect to their importance in the evolution of morphological diversity. Several biologists, including Sean B. Carroll suggest that "changes in the cis-regulatory systems of genes" are more significant than "changes in gene number or protein function". These researchers argue that the combinatorial nature of transcriptional regulation allows a rich substrate for morphological diversity, since variations in the level, pattern, or timing of gene expression may provide more variation for natural selection to act upon than changes in the gene product alone.

Carroll, reviewing a quarter-century of work on the genetic theory of morphological evolution, proposed 8 principles towards such a theory, namely that form (body plan) largely evolves by changing the expression of highly conserved proteins, and that those changes happen mainly through mutations in the cis-regulatory sequences which regulate large networks of genes via pleiotropic developmental loci. Carroll's 8 principles are:

1. Mosaic pleiotropy: most of the proteins that control development take part in many independent developmental processes and give pattern to many dissimilar body structures.

2. Ancestral genetic complexity: Animal taxa with disparate morphology, like cnidarians, insects, and vertebrates and most other phyla, share similar body-constructing genetic toolkits and body-patterning genes, such as Hox.

3. Functional equivalence: Many of the homologous toolkit proteins (orthologs and paralogs) from dissimilar animals (like insects and vertebrates) can be substituted for each other and continue to function. The proteins and their interactions (e.g. with receptors) have therefore changed little in the past billion years.

4. Deep homology: The development of structures like eyes, limbs, and hearts (in phyla as diverse as insects and vertebrates, so the structures were considered non-homologous) are controlled by similar sets of genes and tightly conserved genetic regulatory circuits. For example, PAX-6 governs eye development in the whole animal kingdom, and distal-less governs appendage development in different phyla. This has prompted a revision of the concept of homology.

5. Infrequent toolkit gene duplication: Toolkit genes are much more rarely duplicated than genes in other gene families. Toolkit gene duplication may be selected against, since some developmental processes are sensitive to gene dosage.

6. Heterotopy: Morphological evolution is accompanied by changes in the spatial regulation of toolkit genes during development, along with the genes that they regulate. For example, the spatial location of a developmental process like the formation of a limb or of a pigmentation pattern is modified between species.

7. Modularity of cis-regulatory elements: Pleiotropic toolkit loci are distinguished by big, complicated, and modular cis-regulatory elements. For example, while a non-pleiotropic rhodopsin gene in *Drosophila* has a CRE just a few hundred base pairs long, the pleiotropic eyeless cis-regulatory region contains 6 CREs in over 7000 base pairs.

8. Vast regulatory networks: Each regulatory protein controls "scores to hundreds" of cis-regulatory elements. For instance, 67 *Drosophila* transcription factors controlled on average 124 target genes each.

Consolidation of Epigenetic Changes

Epigenetic alterations of gene regulation or phenotype generation that are subsequently consolidated by changes at the gene level constitute another class of mechanisms for evolutionary innovation. Epigenetic changes include modification of the genetic material due to methylation and other reversible chemical alteration, as well as nonprogrammed remolding of the organism by physical and other environmental effects due to the inherent plasticity of developmental mechanisms. The biologists Stuart A. Newman and Gerd B. Müller have suggested that organisms early in the history of multicellular life were more susceptible to this second category of epigenetic determination than are modern organisms, providing a basis for early macroevolutionary changes.

Ecological Evolutionary Developmental Biology

Ecological evolutionary developmental biology (Eco-evo-devo) is a field that integrates research from developmental biology and ecology to examine their relationship with evolutionary theory. Researchers in this field study concepts and mechanisms such as developmental plasticity, epigenetic inheritance, genetic assimilation, niche construction and symbiosis.

Key Questions

Armin Moczek and 22 colleagues, reviewing the whole field of evo-devo in 2015, note that it unifies biological disciplines including evolution, development, paleontology, neurobiology, cellular biology, molecular biology, quantitative genetics, the study of human diseases, and ecology. They identify questions at the heart of all these disciplines that evo-devo should be able to answer, and note that it should also be able to inform decisions on policy and to improve science education.

In the view of Moczek and colleagues, evo-devo has already transformed understanding of 4 areas of evolutionary biology:

1. The origins of novelty: Darwinism is based on descent with modification, i.e. building on the old, so the new requires explanation.

2. The causes of variation: Darwinism identified variation as the material for evolution, but assumed mutation was its source. Evo-devo shows that several other sources, namely: duplication of genes and body parts; modification of gene control domains; interactions within gene regulatory networks; and co-option at different levels, all enable new traits to evolve.

3. The sources of homology: In 1843, Richard Owen defined homology as "the same organ in different animals under every variety of form and function". Evo-devo enables sameness to be defined more precisely; in 2009, N. Shubin and colleagues including Carroll identified "deep homology" of developmental mechanisms.

4. Convergent evolution of phenotypes: The same traits can evolve repeatedly by a variety of mechanisms, such as selection on the same gene loci, repeated co-option of particular genetic and developmental modules into new places, or reawakening of dormant developmental pathways.

Evolution has taken specific paths, recorded partly in the genomes of different species, the evo-devo mechanisms showing how change can occur, and partly in the fossil record, the palaeontological evidence showing what in fact happened. Evo-devo can thus clarify palaeontology, and has already done so in cases like the explanation of the origins of the complicated shell of turtles, unique to this group of reptiles.

References

- West-Eberhard, M-J. (2003). Developmental plasticity and evolution. New York: Oxford University Press. ISBN 978-0-19-512235-0.

- Gould, Stephen Jay (1977). Ontogeny and Phylogeny. Cambridge, Massachusetts: Harvard University Press. ISBN 0-674-63940-5.

- Carroll, Sean B.; Grenier, Jennifer K.; Weatherbee, Scott D. (2005). From DNA to Diversity: Molecular Genetics and the Evolution of Animal Design — Second Edition. Blackwell Publishing. ISBN 1-4051-1950-0.

- Jablonka, Eva; Lamb, Marion (1995). Epigenetic Inheritance and Evolution: The Lamarckian Dimension. Oxford, New York: Oxford University Press. ISBN 978-0-19-854063-2.

Branches of Evolutionary Biology

Evolutionary biology is an interdisciplinary subject. The genetic difference within and between populations is population genetics while evolutionary biology studies the evolutionary process that has produced the diversity of life. This chapter is a compilation of population genetics and evolutionary ecology.

Population Genetics

Population genetics is the study of the distribution and change in frequency of alleles within populations, and as such it sits firmly within the field of evolutionary biology. The main processes of evolution are natural selection, genetic drift, gene flow and mutation and they form an integral part of the theory that underpins population genetics. Studies in this branch of biology examine such phenomena as adaptation, speciation, population subdivision, and population structure.

Population genetics was a vital ingredient in the emergence of the modern evolutionary synthesis. Its primary founders were Sewall Wright, J. B. S. Haldane and Ronald Fisher, who also laid the foundations for the related discipline of quantitative genetics.

Traditionally a highly mathematical discipline, modern population genetics encompasses theoretical, lab, and field work. Computational approaches, often utilising coalescent theory, have played a central role since the 1980s. In the laboratory, the use of DNA fingerprinting tools have also played a central role since the 1980s.

History

Population genetics began as a reconciliation of Mendelian inheritance and biostatistics models. A key step was the work of the British biologist and statistician Ronald Fisher. In a series of papers starting in 1918 and culminating in his 1930 book *The Genetical Theory of Natural Selection*, Fisher showed that the continuous variation measured by the biometricians could be produced by the combined action of many discrete genes, and that natural selection could change allele frequencies in a population, resulting in evolution. In a series of papers beginning in 1924, another British geneticist, J.B.S. Haldane worked out the mathematics of allele frequency change at a single gene locus under a broad range of conditions. Haldane also applied statistical analysis to real-world examples of natural selection, such as the Peppered moth evolution and industrial melanism, and showed that selection coefficients could be larger than Fisher assumed, leading to more rapid adaptive evolution.

The American biologist Sewall Wright, who had a background in animal breeding experiments, focused on combinations of interacting genes, and the effects of inbreeding on small, relatively isolated populations that exhibited genetic drift. In 1932, Wright introduced the concept of an adaptive landscape and argued that genetic drift and inbreeding could drive a small, isolated sub-population away from an adaptive peak, allowing natural selection to drive it towards different adaptive peaks.

The work of Fisher, Haldane and Wright founded the discipline of population genetics. This integrated natural selection with Mendelian genetics, which was the critical first step in developing a unified theory of how evolution worked. John Maynard Smith was Haldane's pupil, whilst W.D. Hamilton was heavily influenced by the writings of Fisher. The American George R. Price worked with both Hamilton and Maynard Smith. American Richard Lewontin and Japanese Motoo Kimura were heavily influenced by Wright.

Modern Evolutionary Synthesis

The mathematics of population genetics were originally developed as the beginning of the modern evolutionary synthesis. According to Beatty (1986), population genetics defines the core of the modern synthesis. In the first few decades of the 20th century, most field naturalists continued to believe that Lamarckian and orthogenic mechanisms of evolution provided the best explanation for the complexity they observed in the living world. However, as the field of genetics continued to develop, those views became less tenable. During the modern evolutionary synthesis, these ideas were purged, and only evolutionary causes that could be expressed in the mathematical framework of population genetics were retained. Consensus was reached as to which evolutionary factors might influence evolution, but not as to the relative importance of the various factors.

Theodosius Dobzhansky, a postdoctoral worker in T. H. Morgan's lab, had been influenced by the work on genetic diversity by Russian geneticists such as Sergei Chetverikov. He helped to bridge the divide between the foundations of microevolution developed by the population geneticists and the patterns of macroevolution observed by field biologists, with his 1937 book *Genetics and the Origin of Species*. Dobzhansky examined the genetic diversity of wild populations and showed that, contrary to the assumptions of the population geneticists, these populations had large amounts of genetic diversity, with marked differences between sub-populations. The book also took the highly mathematical work of the population geneticists and put it into a more accessible form. Many more biologists were influenced by population genetics via Dobzhansky than were able to read the highly mathematical works in the original.

Selection vs. Genetic Drift

Fisher and Wright had some fundamental disagreements about the relative roles of selection and drift.

In Great Britain E.B. Ford, the pioneer of ecological genetics, continued throughout the 1930s and 1940s to demonstrate the power of selection due to ecological factors including the ability to maintain genetic diversity through genetic polymorphisms such as human blood types. Ford's work, in collaboration with Fisher, contributed to a shift in emphasis during the course of the modern synthesis towards natural selection over genetic drift.

Recent studies of eukaryotic transposable elements, and of their impact on speciation, point again to a major role of nonadaptive processes such as mutation and genetic drift. Mutation and genetic drift are also viewed as major factors in the evolution of genome complexity.

Fundamentals

Biston betularia f. typica is the white-bodied form of the peppered moth.

Biston betularia f. carbonaria is the black-bodied form of the peppered moth.

Population genetics is the study of the frequency and interaction of alleles and genes in populations. A sexual population is a set of organisms in which any pair of members can breed freely together. This implies that all members belong to the same species and are located near each other.

For example, all of the moths of the same species living in an isolated forest are a population. A gene in this population may have several alternate forms, which account for variations between the phenotypes of the organisms. An example might be a gene for coloration in moths that has two alleles: black and white. A gene pool is the complete set of alleles for a gene in a single population; the allele frequency for an allele is the fraction of the genes in the pool that is composed of that allele (for example, what fraction of moth coloration genes are the black allele). Evolution occurs when there are changes in the frequencies of alleles within a population; for example, the allele for black color in a population of moths becoming more common.

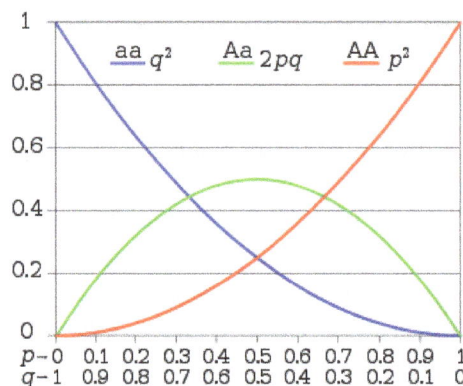

Hardy–Weinberg genotype frequencies for two alleles: the horizontal axis shows the two allele frequencies p and q and the vertical axis shows the genotype frequencies. Each curve shows one of the three possible genotypes.

Four Processes

Selection

Natural selection, which includes sexual selection, is the fact that some traits make it more likely for an organism to survive and reproduce. Population genetics describes natural selection by defining fitness as a propensity or probability of survival and reproduction in a particular environment. The fitness is normally given by the symbol **w=1-s** where **s** is the selection coefficient. Natural selection acts on phenotypes, or the observable characteristics of organisms, but the genetically heritable basis of any phenotype which gives a reproductive advantage will become more common in a population. In this way, natural selection converts differences in fitness into changes in allele frequency in a population over successive generations.

Before the advent of population genetics, many biologists doubted that small differences in fitness were sufficient to make a large difference to evolution. Population geneticists addressed this concern in part by comparing selection to genetic drift. Selection can overcome genetic drift when **s** is greater than 1 divided by the effective population size. When this criterion is met, the probability that a new advantageous mutant becomes fixed is approximately equal to 2s. The time until fixation of such an allele depends little on genetic drift, and is approximately proportional to log(sN)/s.

Hardy–Weinberg Principle

Natural selection will only cause evolution if there is enough genetic variation in a population. Before the discovery of Mendelian genetics, one common hypothesis was blending inheritance. But with blending inheritance, genetic variance would be rapidly lost, making evolution by natural or sexual selection implausible. The *Hardy–Weinberg principle* provides the solution to how variation is maintained in a population with Mendelian inheritance. According to this principle, the frequencies of alleles (variations in a gene) will remain constant in the absence of selection, mutation, migration and genetic drift. The Hardy–Weinberg "equilibrium" refers to this stability of allele frequencies over time.

A second component of the Hardy–Weinberg principle concerns the effects of a single generation of random mating. In this case, the genotype frequencies can be predicted from the allele frequencies. For example, in the simplest case of a single locus with two alleles: the dominant allele is denoted A and the recessive a and their frequencies are denoted by p and q; freq(A) = p; freq(a) = q; $p + q = 1$. If the genotype frequencies are in Hardy–Weinberg proportions resulting from random mating, then we will have freq(AA) = p^2 for the AA homozygotes in the population, freq(aa) = q^2 for the **aa** homozygotes, and freq(Aa) = $2pq$ for the heterozygotes.

Genetic Drift

Genetic drift is a change in allele frequencies caused by random sampling. That is, the alleles in the offspring are a random sample of those in the parents. Genetic drift may cause gene variants to disappear completely, and thereby reduce genetic variability. In contrast to natural selection, which makes gene variants more common or less common depending on their reproductive success, the changes due to genetic drift are not driven by environmental or adaptive pressures, and may be beneficial, neutral, or detrimental to reproductive success.

The effect of genetic drift is larger for alleles present in few copies than when an allele is present in many copies. Scientists wage vigorous debates over the relative importance of genetic drift compared with natural selection. Ronald Fisher held the view that genetic drift plays at the most a minor role in evolution, and this remained the dominant view for several decades. In 1968 Motoo Kimura rekindled the debate with his neutral theory of molecular evolution which claims that most of the changes in the genetic material are caused by neutral mutations and genetic drift. The role of genetic drift by means of sampling error in evolution has been criticized by John H Gillespie and Will Provine, who argue that selection on linked sites is a more important stochastic force.

The population genetics of genetic drift are described using either branching processes or a diffusion equation describing changes in allele frequency. These approaches are usually applied to the Wright-Fisher and John Moran models of population genetics. Assuming genetic drift is the only evolutionary force acting on an allele, after t generations in many replicated populations, starting with allele frequencies of p and q, the variance in allele frequency across those populations is

$$V_t \approx pq\left(1-\exp\left\{-\frac{t}{2N_e}\right\}\right).$$

Mutation

Mutation is the ultimate source of genetic variation in the form of new alleles. Mutation can result in several different types of change in DNA sequences; these can either have no effect, alter the product of a gene, or prevent the gene from functioning. Studies in the fly *Drosophila melanogaster* suggest that if a mutation changes a protein produced by a gene, this will probably be harmful, with about 70 percent of these mutations having damaging effects, and the remainder being either neutral or weakly beneficial.

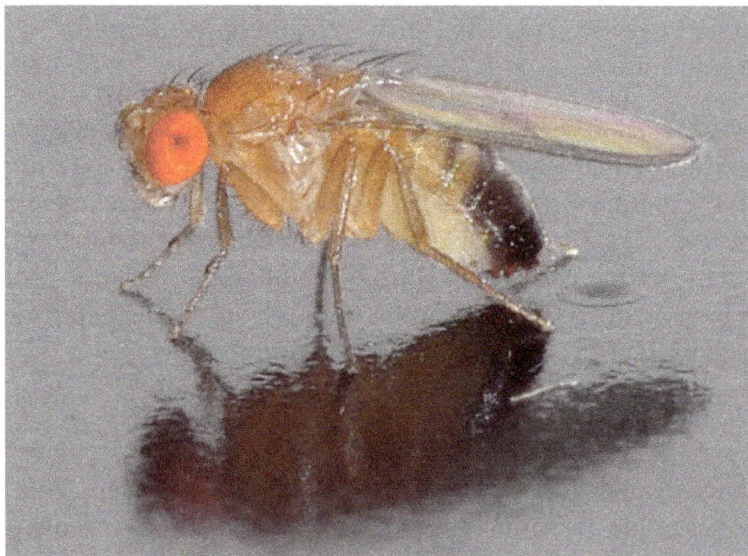

Drosophila melanogaster

Mutations can involve large sections of DNA becoming duplicated, usually through genetic recombination. These duplications are a major source of raw material for evolving new genes, with tens to hundreds of genes duplicated in animal genomes every million years. Most genes belong to larger families of homologous shared ancestry. Novel genes are produced by several methods, commonly through the duplication and mutation of an ancestral gene, or by recombining parts of different genes to form new combinations with new functions. Here, protein domains act as modules, each with a particular and independent function, that can be mixed together to produce genes encoding new proteins with novel properties. For example, the human eye uses four genes to make structures that sense light: three for the cone cell which produce color vision and one for the rod cell which produces night vision; all four arose from a single ancestral gene. Another advantage of duplicating a gene (or even an entire genome) is that this increases redundancy; this allows one gene in the pair to acquire a new function while the other copy performs the original function. Other types of mutation occasionally create new genes from previously noncoding DNA.

In addition to being a major source of variation, mutation may also function as a mechanism of evolution when there are different probabilities at the molecular level for different mutations to occur, a process known as mutation bias. If two genotypes, for example one with the nucleotide G and another with the nucleotide A in the same position, have the same fitness, but mutation from G to A happens more often than mutation from A to G, then genotypes with A will tend to evolve. Different insertion vs. deletion mutation biases in different taxa can lead to the evolution of different genome sizes. Developmental or mutational biases have also been observed in morphological evolution. For example, according to the phenotype-first theory of evolution, mutations can eventually cause the genetic assimilation of traits that were previously induced by the environment.

Mutation bias effects are superimposed on other processes. If selection would favor either one out of two mutations, but there is no extra advantage to having both, then the mutation that occurs the most frequently is the one that is most likely to become fixed in a population. Mutations leading to the loss of function of a gene are much more common than mutations that produce a new, fully functional gene. Most loss of function mutations are selected against. But when selection is weak, mutation bias towards loss of function can affect evolution. For example, pigments are no longer useful when animals live in the darkness of caves, and tend to be lost. This kind of loss of function can occur because of mutation bias, and/or because the function had a cost, and once the benefit of the function disappeared, natural selection leads to the loss. Loss of sporulation ability in a bacterium during laboratory evolution appears to have been caused by mutation bias, rather than natural selection against the cost of maintaining sporulation ability. When there is no selection for loss of function, the speed at which loss evolves depends more on the mutation rate than it does on the effective population size, indicating that it is driven more by mutation bias than by genetic drift.

Evolution of Mutation Rate

Due to the damaging effects that mutations can have on cells, organisms have evolved mechanisms such as DNA repair to remove mutations. Therefore, the optimal mutation rate for a species may be trade-off between costs of a high mutation rate, such as deleterious mutations, and the metabolic costs of maintaining systems to reduce the mutation rate, such as DNA repair enzymes. Viruses that use RNA as their genetic material have rapid mutation rates, which can be an advantage since these viruses will evolve constantly and rapidly, and thus evade the defensive responses of e.g. the human immune system.

Gene Flow

Gene flow is the exchange of genes between populations, which are usually of the same species. Examples of gene flow within a species include the migration and then breeding of organisms, or the exchange of pollen. Gene transfer between species includes the formation of hybrid organisms and horizontal gene transfer.

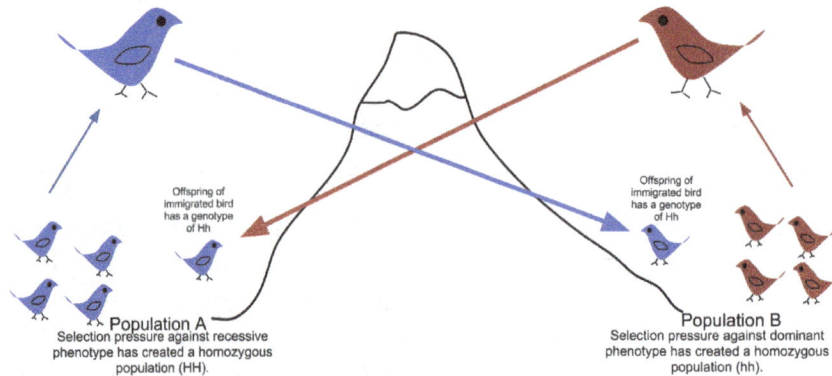

Offspring of immigrated bird has a genotype of Hh

Offspring of immigrated bird has a genotype of Hh

Population A
Selection pressure against recessive phenotype has created a homozygous population (HH).

Population B
Selection pressure against dominant phenotype has created a homozygous population (hh).

Gene flow is the transfer of alleles from one population to another population through immigration of individuals. In this example, one of the birds from population A immigrates to population B, which has fewer of the dominant alleles, and through mating incorporates its alleles into the other population.

Gray tree frog

Migration into or out of a population can change allele frequencies, as well as introducing genetic variation into a population. Immigration may add new genetic material to the established gene pool of a population. Conversely, emigration may remove genetic material. Population genetic models can be used to reconstruct the history of gene flow between populations.

Reproductive Isolation

As barriers to reproduction between two diverging populations are required for the populations to become new species, gene flow may slow this process by spreading genetic differences between the populations. Gene flow is hindered by mountain ranges, oceans and deserts or even man-made structures such as the Great Wall of China, which has hindered the flow of plant genes.

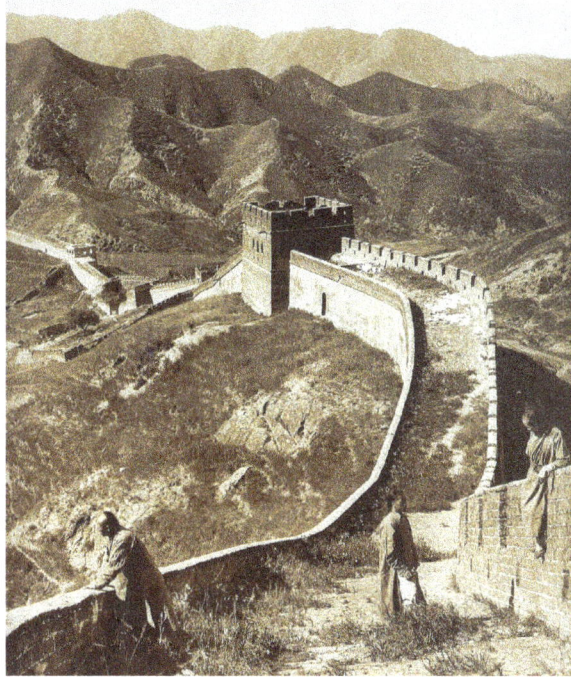

The Great Wall in 1907

Depending on how far two species have diverged since their most recent common ancestor, it may still be possible for them to produce offspring, as with horses and donkeys mating to produce mules. Such hybrids are generally infertile, due to the two different sets of chromosomes being unable to pair up during meiosis. In this case, closely related species may regularly interbreed, but hybrids will be selected against and the species will remain distinct. However, viable hybrids are occasionally formed and these new species can either have properties intermediate between their parent species, or possess a totally new phenotype. The importance of hybridization in creating new species of animals is unclear, although cases have been seen in many types of animals, with the gray tree frog being a particularly well-studied example.

Hybridization is, however, an important means of speciation in plants, since polyploidy (having more than two copies of each chromosome) is tolerated in plants more readily than in animals. Polyploidy is important in hybrids as it allows reproduction, with the two different sets of chromosomes each being able to pair with an identical partner during meiosis. Polyploids also have more genetic diversity, which allows them to avoid inbreeding depression in small populations.

Genetic Structure

Because of physical barriers to migration, along with limited tendency for individuals to move or spread (vagility), and tendency to remain or come back to natal place (philopatry), natural populations rarely all interbreed as convenient in theoretical random models (panmixy) (Buston *et al.*, 2007). There is usually a geographic range within which individuals are more closely related to one another than those randomly selected from the general population. This is described as the extent to which a population is genetically structured (Repaci *et al.*, 2007). Genetic structuring can be caused by migration due to historical climate change, species range expansion or current availability of habitat.

Horizontal Gene Transfer

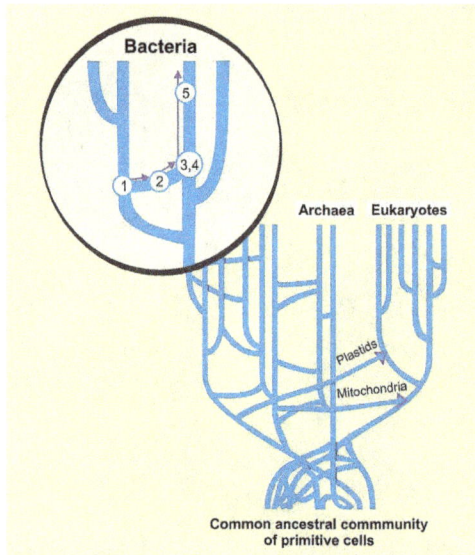

Current tree of life showing vertical and horizontal gene transfers.

Horizontal gene transfer is the transfer of genetic material from one organism to another organism that is not its offspring; this is most common among bacteria. In medicine, this contributes to the spread of antibiotic resistance, as when one bacteria acquires resistance genes it can rapidly transfer them to other species. Horizontal transfer of genes from bacteria to eukaryotes such as the yeast *Saccharomyces cerevisiae* and the adzuki bean beetle *Callosobruchus chinensis* may also have occurred. An example of larger-scale transfers are the eukaryotic bdelloid rotifers, which appear to have received a range of genes from bacteria, fungi, and plants. Viruses can also carry DNA between organisms, allowing transfer of genes even across biological domains. Large-scale gene transfer has also occurred between the ancestors of eukaryotic cells and prokaryotes, during the acquisition of chloroplasts and mitochondria.

Complications

Basic models of population genetics consider only one gene locus at a time. In practice, epistatic and linkage relationships between loci may also be important.

Epistasis

Because of epistasis, the phenotypic effect of an allele at one locus may depend on which alleles are present at many other loci. Selection does not act on a single locus, but on a phenotype that arises through development from a complete genotype.

According to Lewontin (1974), the theoretical task for population genetics is a process in two spaces: a "genotypic space" and a "phenotypic space". The challenge of a *complete* theory of population genetics is to provide a set of laws that predictably map a population of genotypes (G_1) to a phenotype space (P_1), where selection takes place, and another set of laws that map the resulting population (P_2) back to genotype space (G_2) where Mendelian genetics can predict the next generation of genotypes, thus completing the cycle. Even leaving aside for the moment the non-Mendelian aspects of molecular genetics, this is clearly a gargantuan task. Visualizing this transformation schematically:

$$G_1 \xrightarrow{T_1} P_1 \xrightarrow{T_2} P_2 \xrightarrow{T_3} G_2 \xrightarrow{T_4} G_{1'} \rightarrow \cdots$$

(adapted from Lewontin 1974, p. 12). XD

T_1 represents the genetic and epigenetic laws, the aspects of functional biology, or development, that transform a genotype into phenotype. We will refer to this as the "genotype-phenotype map". T_2 is the transformation due to natural selection, T_3 are epigenetic relations that predict genotypes based on the selected phenotypes and finally T_4 the rules of Mendelian genetics.

In practice, there are two bodies of evolutionary theory that exist in parallel, traditional population genetics operating in the genotype space and the biometric theory used in plant and animal breeding, operating in phenotype space. The missing part is the mapping between the genotype and phenotype space. This leads to a "sleight of hand" (as Lewontin terms it) whereby variables in the equations of one domain, are considered parameters or *constants*, where, in a full-treatment they would be transformed themselves by the evolutionary process and are in reality *functions* of the state variables in the other domain. The "sleight of hand" is assuming that we know this mapping. Proceeding as if we do understand it is enough to analyze many cases of interest. For example, if the phenotype is almost one-to-one with genotype (sickle-cell disease) or the time-scale is sufficiently short, the "constants" can be treated as such; however, there are many situations where it is inaccurate.

Linkage

If all genes are in linkage equilibrium, the effect of an allele at one locus can be averaged across the gene pool at other loci. In reality, one allele is frequently found in linkage disequilibrium with genes at other loci, especially with genes located nearby on the same chromosome. Recombination breaks up this linkage disequilibrium too slowly to avoid genetic hitchhiking, where an allele at one locus rises to high frequency because it is linked to an allele under selection at a nearby locus. This is a problem for population genetic models that treat one gene locus at a time. It can, however, be exploited as a method for detecting the action of natural selection via selective sweeps.

In the extreme case of primarily asexual populations, linkage is complete, and different population genetic equations can be derived and solved, which behave quite differently from the sexual case. Most microbes, such as bacteria, are asexual. The population genetics of microorganisms lays the foundations for tracking the origin and evolution of antibiotic resistance and deadly infectious pathogens. Population genetics of microorganisms is also an essential factor for devising strategies for the conservation and better utilization of beneficial microbes (Xu, 2010).

Evolutionary Ecology

Evolutionary ecology lies at the intersection of ecology and evolutionary biology. It approaches the study of ecology in a way that explicitly considers the evolutionary histories of species and the interactions between them. Conversely, it can be seen as an approach to the study of evolution that incorporates an understanding of the interactions between the species under consideration. The main subfields of evolutionary ecology are life history evolution, sociobiology (the evolution of social behavior), the evolution of interspecific relations (cooperation, predator–prey interactions, parasitism, mutualism) and the evolution of biodiversity and of communities.

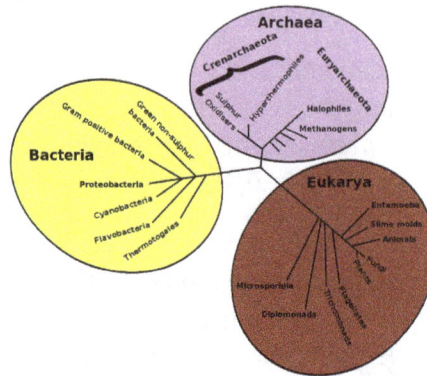

A phylogenetic tree of living things

Pristine, natural environments that have been relatively unaltered by humans are of particular importance in evolutionary ecology because they constitute the environments to which any particular organism has become adapted to over time.

Evolutionary Ecologists

Julia Margaret Cameron's portrait of Darwin

- Charles Darwin, whose theory of natural selection is essential background to understanding evolutionary ecology and had explicitly include population dynamics;

- George Evelyn Hutchinson

- Robert MacArthur

- Eric Pianka

- Michael Rosenzweig

- David Lack

- R. A. Fisher whose 1930 fundamental theorem of natural selection recognised the power of rigorous application of the theory of natural selection to population biology.

- Thierry Lodé

Evolutionary Models

A large part of Evolutionary ecology is about utilising models and finding empirical data as proof. Examples include the Lack clutch size model devised by David Lack; the 1968 model on the specialization of species by Richard Levins; and Law & Diekmann's models on mutualisms.

References

- Mayr, Ernst (1988). Toward a New Philosophy of Biology: Observations of an Evolutionist. Cambridge, MA: Belknap Press of Harvard University Press. p. 402. ISBN 0-674-89665-3.

- Nicholas H. Barton; Derek E. G. Briggs; Jonathan A. Eisen; David B. Goldstein; Nipam H. Patel (2007). Evolution. Cold Spring Harbor Laboratory Press. p. 417. ISBN 0-87969-684-2.

- Sean B. Carroll; Jennifer K. Grenier; Scott D. Weatherbee (2005). From DNA to Diversity: Molecular Genetics and the Evolution of Animal Design. Second Edition. Oxford: Blackwell Publishing. ISBN 1-4051-1950-0.

- West-Eberhard, M-J. (2003). Developmental plasticity and evolution. New York: Oxford University Press. ISBN 978-0-19-512235-0.

- Gregory TR, Hebert PD; Hebert (1999). "The modulation of DNA content: proximate causes and ultimate consequences". Genome Res. 9 (4): 317–24. doi:10.1101/gr.9.4.317 (inactive 2015-01-09). PMID 10207154.

- Scott-Phillips, T. C., Laland, K. N., Shuker, D. M., Dickins, T. E. and West, S. A. (2014). "The Niche Construction Perspective: A Critical Appraisal". Evolution 68: 1231-1243.

- Gravel, S., S. (2012). "Population Genetics Models of Local Ancestry". Genetics. 1202 (2): 4811. arXiv:1202.4811. Bibcode:2012arXiv1202.4811G. doi:10.1534/genetics.112.139808.

- Masel J (2011). "Genetic drift". Current Biology. 21 (20): R837–R838. doi:10.1016/j.cub.2011.08.007. PMID 22032182.

- Wahl L.M. (2011). "Fixation when N and s Vary: Classic Approaches Give Elegant New Results". Genetics. 188 (4): 783–785. doi:10.1534/genetics.111.131748. PMC 3176088. PMID 21828279.

- Orr, H. A. (2010). "The population genetics of beneficial mutations". Philosophical Transactions of the Royal Society B: Biological Sciences. 365 (1544): 1195–1201. doi:10.1098/rstb.2009.0282.

Key Concepts of Evolutionary Biology

The sharing of a recent common ancestor by a group of organisms is known as common descent and adaption is the trait that has been maintained by the organism by the means of natural selection. Speciation and modern evolution synthesis are some other significant key concepts of evolutionary biology. The chapter strategically encompasses and incorporates the major components and key concepts of evolutionary biology.

Common Descent

Descent describes how, in evolutionary biology, a group of organisms share a most recent common ancestor. There is evidence of common descent that all life on Earth is descended from the last universal ancestor. In July 2016, scientists reported identifying a set of 355 genes from the Last Universal Common Ancestor (LUCA) of all organisms living on Earth.

Common ancestry between organisms of different species arises during speciation, in which new species are established from a single ancestral population. Organisms which share a more recent common ancestor are more closely related. The most recent common ancestor of all currently living organisms is the last universal ancestor, which lived about 3.9 billion years ago. The two earliest evidences for life on Earth are graphite found to be biogenic in 3.7 billion-year-old metasedimentary rocks discovered in western Greenland and microbial mat fossils found in 3.48 billion-year-old sandstone discovered in Western Australia. All currently living organisms on Earth share a common genetic heritage (universal common descent), with each being the descendant from a single original species, though the suggestion of substantial horizontal gene transfer during early evolution has led to questions about monophyly of life.

Universal common descent through an evolutionary process was first proposed by the English naturalist Charles Darwin in *On the Origin of Species* (1859), which concluded: "There is grandeur in this view of life, with its several powers, having been originally breathed into a few forms or into one; and that, whilst this planet has gone cycling on according to the fixed law of gravity, from so simple a beginning endless forms most beautiful and most wonderful have been, and are being, evolved."

History

In the 1740s, French mathematician Pierre Louis Maupertuis made the first known suggestion in a series of essays that all organisms may have had a common ancestor, and that they had diverged through random variation and natural selection. In *Essai de cosmologie* (1750), Maupertuis noted:

May we not say that, in the fortuitous combination of the productions of Nature, since only those creatures *could* survive in whose organizations a certain degree of adaptation was present, there is nothing extraordinary in the fact that such adaptation is actually found in all these species which now exist? Chance, one might say, turned out a vast number of individuals; a small proportion of these were organized in such a manner that the animals' organs could satisfy their needs. A much greater number showed neither adaptation nor order; these last have all perished.... Thus the species which we see today are but a small part of all those that a blind destiny has produced.

In 1790, Immanuel Kant wrote in *Kritik der Urteilskraft* (*Critique of Judgement*) that the analogy of animal forms implies a common original type, and thus a common parent.

In 1794, Charles Darwin's grandfather, Erasmus Darwin, asked:

[W]ould it be too bold to imagine, that in the great length of time, since the earth began to exist, perhaps millions of ages before the commencement of the history of mankind, would it be too bold to imagine, that all warm-blooded animals have arisen from one living filament, which THE GREAT FIRST CAUSE endued with animality, with the power of acquiring new parts attended with new propensities, directed by irritations, sensations, volitions, and associations; and thus possessing the faculty of continuing to improve by its own inherent activity, and of delivering down those improvements by generation to its posterity, world without end?

Charles Darwin's views about common descent, as expressed in *On the Origin of Species*, were that it was possible that there was only one progenitor for all life forms:

Therefore I should infer from analogy that probably all the organic beings which have ever lived on this earth have descended from some one primordial form, into which life was first breathed.

Evidence of Universal Common Descent

Common Biochemistry and Genetic Code

All known forms of life are based on the same fundamental biochemical organization: genetic information encoded in DNA, transcribed into RNA, through the effect of protein- and RNA-enzymes, then translated into proteins by (highly similar) ribosomes, with ATP, NADPH and others as energy sources, etc. Furthermore, the genetic code (the "translation table" according to which DNA information is translated into proteins) is nearly identical for all known lifeforms, from bacteria and archaea to animals and plants. The universality of this code is generally regarded by biologists as definitive evidence in favor of the theory of universal common descent. Analysis of the small differences in the genetic code has also provided support for universal common descent. An example would be Cytochrome c which most organisms actually share. A statistical comparison of various alternative hypotheses has shown that universal common ancestry is significantly more probable than models involving multiple origins.

Selectively Neutral Similarities

Similarities which have no adaptive relevance cannot be explained by convergent evolution, and therefore they provide compelling support for the theory of universal common descent.

Such evidence has come from two areas: amino acid sequences and DNA sequences. Proteins with the same three-dimensional structure need not have identical amino acid sequences; any irrelevant similarity between the sequences is evidence for common descent. In certain cases, there are several codons (DNA triplets) that code for the same amino acid. Thus, if two species use the same codon at the same place to specify an amino acid that can be represented by more than one codon, that is evidence for a recent common ancestor.

Other Similarities

The universality of many aspects of cellular life is often pointed to as supportive evidence to the more compelling evidence listed above. These similarities include the energy carrier adenosine triphosphate (ATP), and the fact that all amino acids found in proteins are left-handed. It is, however, possible that these similarities resulted because of the laws of physics and chemistry, rather than universal common descent and therefore resulted in convergent evolution.

Phylogenetic Trees

Phylogenetic Tree of Life

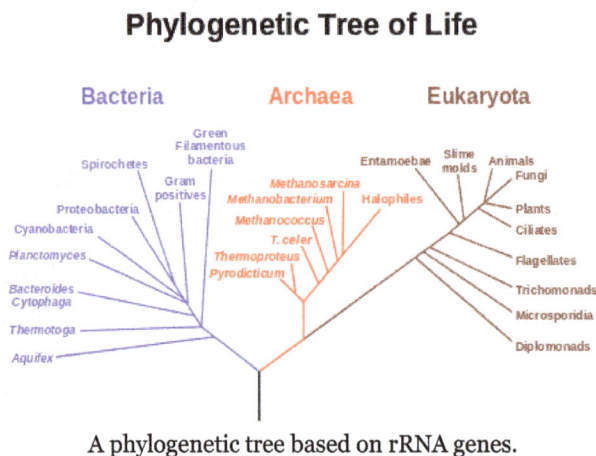

A phylogenetic tree based on rRNA genes.

Another important piece of evidence is that it is possible to construct detailed phylogenetic trees (i.e., "genealogic trees" of species) mapping out the proposed divisions and common ancestors of all living species. In 2010, Douglas L. Theobald published a statistical analysis of available genetic data, mapping them to phylogenetic trees, that gave "strong quantitative support, by a formal test, for the unity of life." It should be noted, however, that the "formal test" was criticised for not including consideration of convergent evolution, and Theobald has defended the method against this claim.

Traditionally, these trees have been built using morphological methods, such as appearance, embryology, etc. Recently, it has been possible to construct these trees using molecular data, based on similarities and differences between genetic and protein sequences. All these methods produce essentially similar results, even though most genetic variation has no influence over external morphology. That phylogenetic trees based on different types of information agree with each other is strong evidence of a real underlying common descent.

Illustrations of Common Descent

Artificial Selection

Artificial selection demonstrates the diversity that can exist among organisms that share a relatively recent common ancestor. In artificial selection, humans selectively direct the breeding of one species at each generation, allowing only those organisms that exhibit desired characteristics to reproduce. These characteristics become increasingly well-developed in successive generations. Artificial selection was successful long before science discovered the genetic basis.

Dog Breeding

The Chihuahua mix and Great Dane both share a common ancestor, the wolf, but show the power of artificial selection to create diversity of form in a relatively short period of time.

The diversity of domesticated dogs is an example of the power of artificial selection. All breeds share common ancestry, having descended from wolves. Humans selectively bred them to enhance specific characteristics, such as color and length or body size. This created a range of breeds that include the Chihuahua, Great Dane, Basset Hound, Pug, and Poodle. Wild wolves, which did not undergo artificial selection, are relatively uniform in comparison.

Wild cabbage

Early farmers cultivated many popular vegetables from the *Brassica oleracea* (wild cabbage) by artificially selecting for certain attributes. Common vegetables such as cabbage, kale, broccoli, cauliflower, kohlrabi and Brussels sprouts are all descendants of the wild cabbage plant. Brussels sprouts were created by artificially selecting for large bud size. Broccoli was bred by selecting for large flower stalks. Cabbage was created by selecting for short petioles. Kale was bred by selecting for large leaves.

Natural Selection

Natural selection is the evolutionary process by which heritable traits that increase an individual's fitness become more common, and heritable traits that decrease an individual's fitness become less common.

Darwin's Finches

During his studies on the Galápagos Islands, Charles Darwin observed 13 species of finches that are closely related and differ most markedly in the shape of their beaks. The beak of each species is suited to the food available in its particular environment, suggesting that beak shapes evolved by natural selection. Large beaks were found on the islands where the primary source of food for the finches are nuts and therefore the large beaks allowed the birds to be better equipped for opening the nuts and staying well nourished. Slender beaks were found on the finches which found insects to be the best source of food on the island they inhabited; their slender beaks allowed the birds to be better equipped for pulling out the insects from their tiny hiding places. The finch is also found on the main land and it is thought that they migrated to the islands and began adapting to their environment through natural selection.

Speciation

Speciation is the evolutionary process by which reproductively isolated biological populations evolve to become distinct species. The biologist Orator F. Cook was the first to coin the term 'speciation' for the splitting of lineages or "cladogenesis," as opposed to "anagenesis" or "phyletic evolution" occurring within lineages. Charles Darwin was the first to describe the role of natural selection in speciation.

There are four geographic modes of speciation in nature, based on the extent to which speciating populations are isolated from one another: allopatric, peripatric, parapatric, and sympatric. Speciation may also be induced artificially, through animal husbandry, agriculture, or laboratory experiments. Whether genetic drift is a minor or major contributor to speciation is the subject matter of much ongoing discussion.

It is widely appreciated that sexual selection could drive speciation in many clades, independently of natural selection. However the term "speciation", in this context, tends to be used in two different, but not mutually exclusive senses. The first and most commonly used sense refers to the "birth" of new species. That is, the splitting of an existing species into two separate species, or the budding off of a new species from a parent species, both driven by a biological "fashion fad" (a preference for a feature, or features, in one or both sexes, that do not necessarily have any adaptive qualities). In the second sense, "speciation" refers the wide-spread tendency of sexual creatures to be grouped into clearly defined species, rather than forming a continuum of phenotypes both in time and space - which would be the more obvious or logical consequence of natural selection. This was indeed recognized by Darwin as problematic, and included in his *On the Origin of Species* (1859), under the heading "Difficulties with the Theory". There are several suggestions as to how mate choice might play a significant role in resolving Darwin's dilemma.

Modes of Speciation

All forms of natural speciation have taken place over the course of evolution; however, debate persists as to the relative importance of each mechanism in driving biodiversity.

One example of natural speciation is the diversity of the three-spined stickleback, a marine fish that, after the last glacial period, has undergone speciation into new freshwater colonies in isolated lakes and streams. Over an estimated 10,000 generations, the sticklebacks show structural differences that are greater than those seen between different genera of fish including variations in fins, changes in the number or size of their bony plates, variable jaw structure, and color differences.

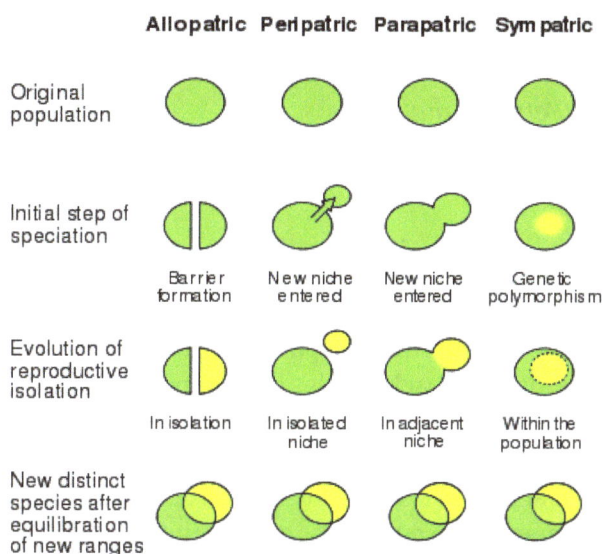

Comparison of allopatric, peripatric, parapatric and sympatric speciation

Allopatric

During allopatric (from the ancient Greek *allos*, "other" + Greek *patrā*, "fatherland") speciation, a population splits into two geographically isolated populations (for example, by habitat fragmentation due to geographical change such as mountain formation). The isolated populations then undergo genotypic and/or phenotypic divergence as: (a) they become subjected to dissimilar selective pressures; (b) they independently undergo genetic drift; (c) different mutations arise in the two populations. When the populations come back into contact, they have evolved such that they are reproductively isolated and are no longer capable of exchanging genes. Island genetics is the term associated with the tendency of small, isolated genetic pools to produce unusual traits. Examples include insular dwarfism and the radical changes among certain famous island chains, for example on Komodo. The Galápagos Islands are particularly famous for their influence on Charles Darwin. During his five weeks there he heard that Galápagos tortoises could be identified by island, and noticed that finches differed from one island to another, but it was only nine months later that he reflected that such facts could show that species were changeable. When he returned to England, his speculation on evolution deepened after experts informed him that these were separate species, not just varieties, and famously that other differing Galápagos birds were all species of finches. Though the finches were less important for Darwin, more recent research has shown the birds now known as Darwin's finches to be a classic case of adaptive evolutionary radiation.

Peripatric

In peripatric speciation, a subform of allopatric speciation, new species are formed in isolated, smaller peripheral populations that are prevented from exchanging genes with the main population. It is related to the concept of a founder effect, since small populations often undergo bottlenecks. Genetic drift is often proposed to play a significant role in peripatric speciation.

Case Studies:

- Mayr bird fauna

- The Australian bird *Petroica multicolor*

- Reproductive isolation occurs in populations of *Drosophila* subject to population bottle-necking

Parapatric

In parapatric speciation, there is only partial separation of the zones of two diverging populations afforded by geography; individuals of each species may come in contact or cross habitats from time to time, but reduced fitness of the heterozygote leads to selection for behaviours or mechanisms that prevent their interbreeding. Parapatric speciation is modelled on continuous variation within a "single," connected habitat acting as a source of natural selection rather than the effects of isolation of habitats produced in peripatric and allopatric speciation.

Parapatric speciation may be associated with differential landscape-dependent selection. Even if there is a gene flow between two populations, strong differential selection may impede assimilation and different species may eventually develop. Habitat differences may be more important in the development of reproductive isolation than the isolation time. Caucasian rock lizards *Darevskia rudis, D. valentini* and *D. portschinskii* all hybridize with each other in their hybrid zone; however, hybridization is stronger between *D. portschinskii* and *D. rudis*, which separated earlier but live in similar habitats than between *D. valentini* and two other species, which separated later but live in climatically different habitats.

Ecologists refer to parapatric and peripatric speciation in terms of ecological niches. A niche must be available in order for a new species to be successful.

Case Studies:

- Ring species

    o The *Larus* gulls form a ring species around the North Pole.

    o The *Ensatina* salamanders, which form a ring round the Central Valley in California.

    o The greenish warbler (*Phylloscopus trochiloides*), around the Himalayas.

- The grass *Anthoxanthum* has been known to undergo parapatric speciation in such cases as mine contamination of an area.

Sympatric

Sympatric speciation refers to the formation of two or more descendant species from a single ancestral species all occupying the same geographic location.

Japanese rice fish

Freshwater angelfish, a cichlid

Often-cited examples of sympatric speciation are found in insects that become dependent on different host plants in the same area. However, the existence of sympatric speciation as a mechanism of speciation remains highly debated.

The best illustrated example of sympatric speciation is that of the cichlids of East Africa inhabiting the Rift Valley lakes, particularly Lake Victoria, Lake Malawi and Lake Tanganyika. There are over 800 described species, and according to estimate, there could be well over 1,600 species in the region. All the species have diversified from a common ancestral fish, the Japanese rice fish (*Oryzias latipes*) about 113 million years ago. Their evolution is cited as an example of both natural and sexual selection. A 2008 study suggests that sympatric speciation has occurred in Tennessee cave salamanders. Sympatric speciation driven by ecological factors may also account for the extraordinary diversity of crustaceans living in the depths of Siberia's Lake Baikal.

Budding speciation has been proposed as a particular form of sympatric speciation, whereby small groups of individuals become progressively more isolated from the ancestral stock by breeding preferentially with one another. This type of speciation would be driven by the conjunction of various advantages of inbreeding such as the expression of advantageous recessive phenotypes, reducing the recombination load, and reducing the cost of sex.

Example: Hawthorn fly

One example of evolution at work is the case of the hawthorn fly (*Rhagoletis pomonella*), also known as the apple maggot fly, which appears to be undergoing sympatric speciation. Different populations of hawthorn fly feed on different fruits. A distinct population emerged in North America in the 19th century some time after apples, a non-native species, were introduced. This apple-feeding population normally feeds only on apples and not on the historically preferred fruit of hawthorns. The current hawthorn feeding population does not normally feed on apples. Some evidence, such as the fact that six out of thirteen allozyme loci are different, that hawthorn flies mature later in the season and take longer to mature than apple flies; and that there is little evidence of interbreeding (researchers have documented a 4-6% hybridization rate) suggests that sympatric speciation is occurring. The emergence of the new hawthorn fly is an example of evolution in progress.

Rhagoletis pomonella

Example: Three-spined Sticklebacks

The three-spined stickleback (*Gasterosteus aculeatus*)

Freshwater three-spined sticklebacks, which have been studied by Dolph Schluter, were once thought to provide an intriguing example best explained by sympatric speciation. Schluter and colleagues found two different species of three-spined sticklebacks in each of five different lakes. Each lake contained a large benthic species with a large mouth that feeds on large prey in the littoral zone, as well as a smaller limnetic species with a smaller mouth that feeds on the small plankton in open water. DNA analysis indicated that each lake was colonized independently, presumably by a marine ancestor, after the last glacial period. It also showed that the two species in each lake are more closely related to each other than they are to any of the species in the other lakes.The two species in each lake are reproductively isolated; neither mates with the other. However, aquarium tests showed that the benthic species from one lake is able to mate with the benthic species from the other lakes. Likewise the limnetic species from the different lakes are able to mate with each other. These benthic and limnetic species even display their mating preferences when presented with sticklebacks from Japanese lakes. A Canadian benthic prefers a Japanese benthic over its close limnetic relative from its own lake.

The researchers concluded that in each lake there had been great competition within a single original species for limited resources. This led to disruptive selection—competition favoring fishes at either extreme of body size and mouth size over those nearer the mean, as well as assortative mating—each size preferred mates like it. The result was a divergence into two subpopulations exploiting different food in different parts of the lake.The fact that this pattern of speciation occurred the same way on three separate occasions suggests strongly that ecological factors in a sympatric population can cause speciation.

However, the DNA evidence cited above is from mitochondrial DNA (mtDNA), which can often move easily between closely related species ("introgression") when they hybridize or engage in despeciation. A more recent study, using genetic markers from the nuclear genome, shows that limnetic forms in different lakes are more closely related to each other (and to marine lineages) than to benthic forms in the same lake. The three-spine stickleback is now usually considered an example of "double invasion" (a form of allopatric speciation) in which repeated invasions of marine forms have subsequently differentiated into benthic and limnetic forms. The three-spine stickleback provides an example of how molecular biogeographic studies that rely solely on mtDNA can be misleading, and that consideration of the genealogical history of alleles from multiple unlinked markers (i.e., nuclear genes) is necessary to infer speciation histories.

Reinforcement

Reinforcement, also called the *Wallace effect*, is the process by which natural selection increases reproductive isolation. It may occur after two populations of the same species are separated and then come back into contact. If their reproductive isolation was complete, then they will have already developed into two separate incompatible species. If their reproductive isolation is incomplete, then further mating between the populations will produce hybrids, which may or may not be fertile. If the hybrids are infertile, or fertile but less fit than their ancestors, then there will be further reproductive isolation and speciation has essentially occurred (e.g., as in horses and donkeys.)

The reasoning behind this is that if the parents of the hybrid offspring each have naturally selected traits for their own certain environments, the hybrid offspring will bear traits from both, therefore would not fit either ecological niche as well as either parent. The low fitness of the hybrids would cause selection to favor assortative mating, which would control hybridization. This is sometimes called the Wallace effect after the evolutionary biologist Alfred Russel Wallace who suggested in the late 19th century that it might be an important factor in speciation. Conversely, if the hybrid offspring are more fit than their ancestors, then the populations will merge back into the same species within the area they are in contact.

Reinforcement favoring reproductive isolation is required for both parapatric and sympatric speciation. Without reinforcement, the geographic area of contact between different forms of the same species, called their "hybrid zone," will not develop into a boundary between the different species. Hybrid zones are regions where diverged populations meet and interbreed. Hybrid offspring are very common in these regions, which are usually created by diverged species coming into secondary contact. Without reinforcement, the two species would have uncontrollable inbreeding. Reinforcement may be induced in artificial selection experiments as described below.

Ecological and Parallel Speciation

Ecological selection is "the interaction of individuals with their environment during resource acquisition". Natural selection is inherently involved in the process of speciation, whereby, "under ecological speciation, populations in different environments, or populations exploiting different resources, experience contrasting natural selection pressures on the traits that directly or indirectly bring about the evolution of reproductive isolation". Evidence for the role ecology plays in the process of speciation exists. Studies of stickleback populations support ecologically-linked speciation arising as a by-product, alongside numerous studies of parallel speciation.

Parallel speciation is where "greater reproductive isolation repeatedly evolves between independent populations adapting to contrasting environments than between independent populations adapting to similar environments". It is established that ecological speciation occurs and with much of the evidence, "...accumulated from top-down studies of adaptation and reproductive isolation".

Artificial Speciation

European mouflon (*Ovis aries musimon*)

New species have been created by domesticated animal husbandry, but the initial dates and methods of the initiation of such species are not clear. Often, the domestic counterpart of the wild ancestor can still interbreed and produce fertile offspring as in the case of domestic cattle, that can be considered the same species as several varieties of wild ox, gaur, yak, etc., or domestic sheep that can interbreed with the mouflon.

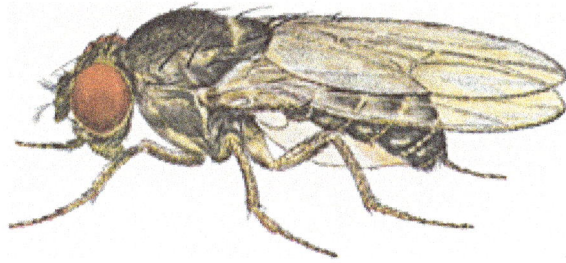

Male *Drosophila pseudoobscura*

The best-documented creations of new species in the laboratory were performed in the late 1980s. William R. Rice and George W. Salt bred *Drosophila melanogaster* fruit flies using a maze with three different choices of habitat such as light/dark and wet/dry. Each generation was placed into the maze, and the groups of flies that came out of two of the eight exits were set apart to breed with each other in their respective groups. After thirty-five generations, the two groups and their offspring were isolated reproductively because of their strong habitat preferences: they mated only within the areas they preferred, and so did not mate with flies that preferred the other areas. The history of such attempts is described by Rice and Elen E. Hostert (1993).

Diane Dodd used a laboratory experiment to show how reproductive isolation can evolve in *Drosophila pseudoobscura* fruit flies after several generations by placing them in different media, starch- and maltose-based media.

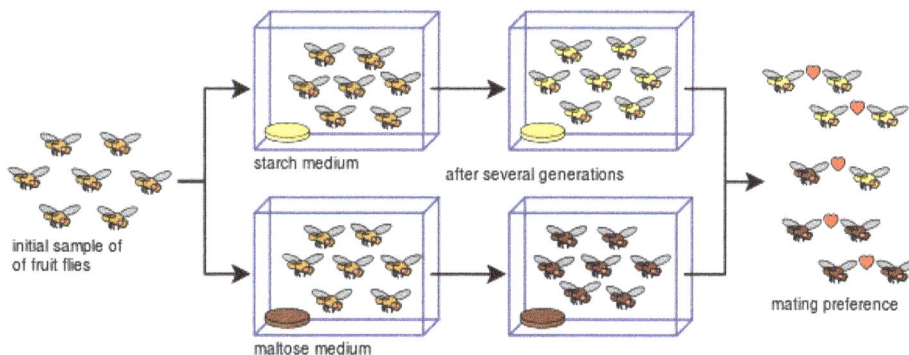

Dodd's experiment has been easy for many others to replicate, including with other kinds of fruit flies and foods. Research in 2005 has shown that this rapid evolution of reproductive isolation may in fact be a relic of infection by *Wolbachia* bacteria.

Alternatively, these observations are consistent with the notion that sexual creatures are inherently reluctant to mate with individuals whose appearance or behavior is different from the norm. The risk that such deviations are due to heritable maladaptations is very high. Thus, if a sexual creature, unable to predict natural selection's future direction, is conditioned to produce the fittest offspring possible, it will avoid mates with unusual habits or features. Sexual creatures will then inevitably tend to group themselves into reproductively isolated species.

Genetics

Few speciation genes have been found. They usually involve the reinforcement process of late stages of speciation. In 2008, a speciation gene causing reproductive isolation was reported. It causes hybrid sterility between related subspecies. The order of speciation of three groups from a common ancestor may be unclear or unknown; a collection of three such species is referred to as a "trichotomy."

Speciation via Polyploidization

Polyploidy is a mechanism that has caused many rapid speciation events in sympatry because offspring of, for example, tetraploid x diploid matings often result in triploid sterile progeny. However, not all polyploids are reproductively isolated from their parental plants, and gene flow may still occur for example through triploid hybrid x diploid matings that produce tetraploids, or matings between meiotically unreduced gametes from diploids and gametes from tetraploids.

It has been suggested that many of the existing plant and most animal species have undergone an event of polyploidization in their evolutionary history. Reproduction of successful polyploid species is sometimes asexual, by parthenogenesis or apomixis, as for unknown reasons many asexual organisms are polyploid. Rare instances of polyploid mammals are known, but most often result in prenatal death.

Hybrid Speciation

Hybridization between two different species sometimes leads to a distinct phenotype. This phenotype can also be fitter than the parental lineage and as such natural selection may then favor these individuals. Eventually, if reproductive isolation is achieved, it may lead to a separate species. However, reproductive isolation between hybrids and their parents is particularly difficult to achieve and thus hybrid speciation is considered an extremely rare event. The mariana mallard is thought to have arisen from hybrid speciation.

Hybridization is an important means of speciation in plants, since polyploidy (having more than two copies of each chromosome) is tolerated in plants more readily than in animals. Polyploidy is important in hybrids as it allows reproduction, with the two different sets of chromosomes each being able to pair with an identical partner during meiosis. Polyploids also have more genetic diversity, which allows them to avoid inbreeding depression in small populations.

Hybridization without change in chromosome number is called homoploid hybrid speciation. It is considered very rare but has been shown in *Heliconius* butterflies and sunflowers. Polyploid speciation, which involves changes in chromosome number, is a more common phenomenon, especially in plant species.

Gene Transposition

Theodosius Dobzhansky, who studied fruit flies in the early days of genetic research in 1930s, speculated that parts of chromosomes that switch from one location to another might cause a species to split into two different species. He mapped out how it might be possible for sections of chromosomes to relocate themselves in a genome. Those mobile sections can cause sterility in inter-species hybrids, which can act as a speciation pressure. In theory, his idea was sound, but scientists long debated whether it actually happened in nature. Eventually a competing theory involving the gradual accumulation of mutations was shown to occur in nature so often that geneticists largely dismissed the moving gene hypothesis.

However, 2006 research shows that jumping of a gene from one chromosome to another can contribute to the birth of new species. This validates the reproductive isolation mechanism, a key component of speciation.

Historical Background

Life Timeline

In addressing the question of the origin of species, there are two key issues: (1) what are the evolutionary mechanisms of speciation, and (2) what accounts for the separateness and individuality of species in the biota? Since Charles Darwin's time, efforts to understand the nature of species have primarily focused on the first aspect, and it is now widely agreed that the critical factor behind the origin of new species is reproductive isolation. Next we focus on the second aspect of the origin of species.

Darwin's Dilemma: Why Do Species Exist?

In *On the Origin of Species* (1859), Darwin interpreted biological evolution in terms of natural selection, but was perplexed by the clustering of organisms into species. Chapter 6 of Darwin's book is entitled "Difficulties of the Theory." In discussing these "difficulties" he noted "Firstly, why, if species have descended from other species by insensibly fine gradations, do we not everywhere see innumerable transitional forms? Why is not all nature in confusion instead of the species being, as we see them, well defined?" This dilemma can be referred to as the absence or rarity of transitional varieties in habitat space.

Another dilemma, related to the first one, is the absence or rarity of transitional varieties in time. Darwin pointed out that by the theory of natural selection "innumerable transitional forms must have existed," and wondered "why do we not find them embedded in countless numbers in the crust of the earth." That clearly defined species actually do exist in nature in both space and time implies that some fundamental feature of natural selection operates to generate and maintain species.

The Effect of Sexual Reproduction on Species Formation

It has been argued that the resolution of Darwin's dilemmas lies in the fact that out-crossing sexual reproduction has an intrinsic cost of rarity. The cost of rarity arises as follows. If, on a resource gradient, a large number of separate species evolve, each exquisitely adapted to a very narrow band on that gradient, each species will, of necessity, consist of very few members. Finding a mate under these circumstances may present difficulties when many of the individuals in the neighborhood belong to other species. Under these circumstances, any species that happens, by chance, to increase in population size (at the expense of one or other of its neighboring species, if the environment is saturated), this will immediately make it easier for its members to find sexual partners. The members of the neighboring species, whose population sizes have decreased, will experience greater difficulty in finding mates, and therefore form pairs less frequently than in the larger species. This has a snowball effect, with large species growing at the expense of the smaller, rarer species, eventually driving them to extinction. Eventually, only a few species remain, each distinctly different from the other. The cost of rarity not only involves the costs of failure to find a mate, but also indirect costs such as the cost of communication in seeking out a partner at low population densities.

A photograph of an African pygmy kingfisher, showing details of appearance and coloration that are shared by adults of that species to a high degree of fidelity.

Bernstein *et al.* argue furthermore that if an environmental gradient is populated by a single species which is perfectly adapted to only a small portion of that environment, it will be difficult for better-adapted individuals to pass their adaptation on to others in region. Such advantageous characteristics are unlikely to be due to a single altered gene, but rather to a combination of several altered genes, each of which, on its own, imparts little or no benefit to its carrier. If such an individual mates with a randomly selected mate, the advantageous combination of genes will be broken up, and the advantage lost, unless it happens to mate with another individual with the same advantageous combination of altered genes. This will be an exceptionally rare event, the consequence of which is that the species will be resistant to change over time or to the budding off of new species.

Rarity brings with it other costs. A rare or unusual feature is very seldom advantageous. In most instances, it will be indicative of a (non-silent) mutation, which is almost certain to be deleterious. It therefore behooves sexual creatures to avoid mates sporting rare or unusual features. Should this be the case, then sexual populations will rapidly shed rare or peripheral phenotypic features, thus canalizing the entire external appearance, as illustrated in the accompanying illustration of the African pygmy kingfisher, *Ispidina picta*. This remarkable uniformity of all the adult members of a sexual species has stimulated the proliferation of field guides on birds, mammals, reptiles, insects, and many other taxons, in which each species can be described in full by means of a single illustration (or a pair of illustrations if there is sufficient sexual dimorphism). Once a population has become as homogeneous in appearance as is typical of most species (and is illustrated in the photograph of the African pygmy kingfisher), its members will avoid mating with members of other populations that look different from themselves. Thus, the avoidance of mates displaying rare and unusual phenotypic features inevitably leads to reproductive isolation, one of the hallmarks of speciation.

In the contrasting case of organisms that reproduce asexually, there is no cost of rarity; consequently, there are only benefits to fine-scale adaptation. Thus, asexual organisms very frequently show the continuous variation in form (often in many different directions) that Darwin expected evolution to produce, making their classification into "species" (more correctly, "morphospecies") very difficult.

Rates of Speciation

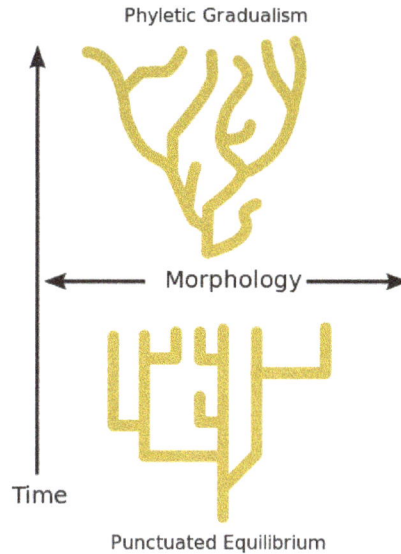

Phyletic Gradualism

Morphology

Time

Punctuated Equilibrium

Phyletic gradualism, above, consists of relatively slow change over geological time. Punctuated equilibrium, bottom, consists of morphological stability and rare, relatively rapid bursts of evolutionary change.

There is debate as to the rate at which speciation events occur over geologic time. While some evolutionary biologists claim that speciation events have remained relatively constant and gradual over time, some palaeontologists such as Niles Eldredge and Stephen Jay Gould have argued that species usually remain unchanged over long stretches of time, and that speciation occurs only over relatively brief intervals, a view known as *punctuated equilibrium.*)

Punctuated Evolution

Evolution can be extremely rapid, as shown in the creation of domesticated animals and plants in a very short geological space of time, spanning only a few tens of thousands of years. Maize (*Zea mays*), for instance, was created in Mexico in only a few thousand years, starting about 7,000 to 12,000 years ago. This raises the question of why the long term rate of evolution is far slower than is theoretically possible.

Plants and Domestic Animals can Differ Markedly from their Wild Ancestors

Top: wild teosinte; middle: maize-teosinte hybrid; bottom: maize

Ancestral wild cabbage	Domesticated cauliflowe
Ancestral Prussian carp	Domestic goldfish
Ancestral mouflon	Domestic sheep

Evolution is imposed on species or groups. It is not planned or striven for in some Lamarckist way. The mutations on which the process depends are random events, and, except for the "silent mutations" which do not affect the functionality or appearance of the carrier, are thus usually disadvantageous, and their chance of proving to be useful in the future is vanishingly small. Therefore, while a species or group might benefit from being able to adapt to a new environment by accumulating a wide range of genetic variation, this is to the detriment of the *individuals* who have to carry these mutations until a small, unpredictable minority of them ultimately contributes to such an adaptation. Thus, the *capability* to evolve is a group adaptation, a concept discredited by (for example) George C. Williams, John Maynard Smith and Richard Dawkins as selectively disadvantageous to the individual.

If sexual creatures avoid mutant mates with strange or unusual characteristics, then mutations that affect the external appearance of their carriers would seldom be passed on to the next and subsequent generations. They would therefore seldom be tested by natural selection. Evolution is, therefore, effectively halted or slowed down considerably. The only mutations that can accumulate in a population, on this punctuated equilibrium view, are ones that have no noticeable effect on the outward appearance and functionality of their bearers (i.e., they are "silent" or "neutral mutations," which can be, and are, used to trace the relatedness and age of populations and species.)

This argument implies that evolution can only occur if mutant mates cannot be avoided, as a result of a severe scarcity of potential mates. This is most likely to occur in small, isolated communities. These occur most commonly on small islands, in remote valleys, lakes, river systems, or caves, or during the aftermath of a mass extinction. Under these circumstances, not only is the choice of mates severely restricted but population bottlenecks, founder effects, genetic drift and inbreeding cause rapid, random changes in the isolated population's genetic composition. Furthermore, hybridization with a related species trapped in the same isolate might introduce additional genetic changes. If an isolated population such as this survives its genetic upheavals, and subsequently expands into an unoccupied niche, or into a niche in which it has an advantage over its competitors, a new species, or subspecies, will have come in being. In geological terms this will be an abrupt event. A resumption of avoiding mutant mates will thereafter result, once again, in evolutionary stagnation.

In apparent confirmation of this punctuated equilibrium view of evolution, the fossil record of an evolutionary progression typically consists of species that suddenly appear, and ultimately disappear, in many cases close to a million years later, without any change in external appearance. Graphically, these fossil species are represented by horizontal lines, whose lengths depict how long each of them existed. The horizontality of the lines illustrates the unchanging appearance of each of the fossil species depicted on the graph. During each species' existence new species appear at random intervals, each also lasting many hundreds of thousands of years before disappearing without a change in appearance. The exact relatedness of these concurrent species is generally impossible to determine. This is illustrated in the following diagram depicting the evolution of modern humans from the time that the hominins separated from the line that led to the evolution of our closest living primate relatives, the chimpanzees.

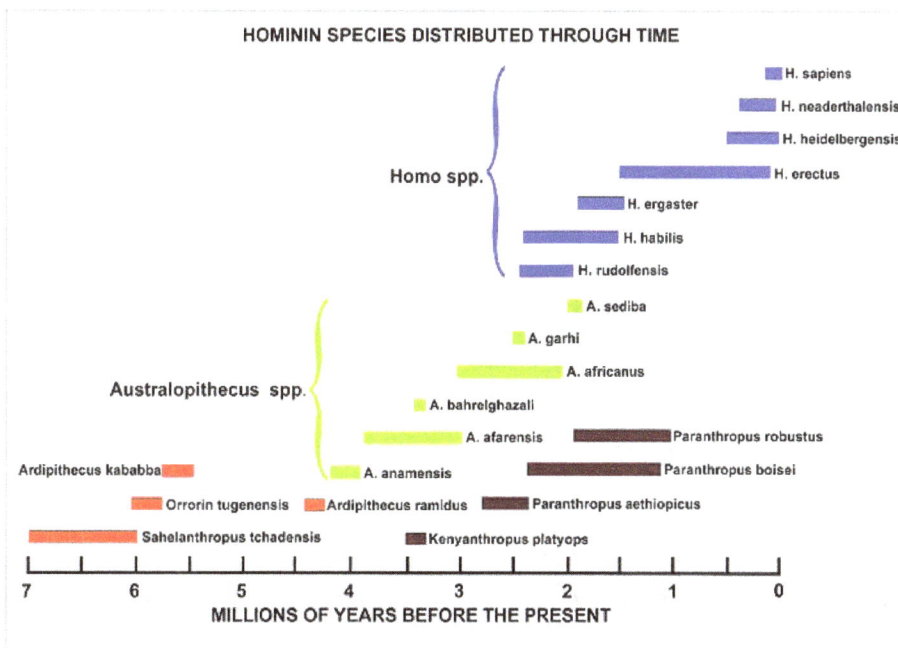

Distribution of Hominin species over time

For similar evolutionary time lines see, for instance, the paleontological list of African dinosaurs, Asian dinosaurs, the Lampriformes and Amiiformes.

Modern Evolutionary Synthesis

The modern evolutionary synthesis (known as the new synthesis, the modern synthesis, the evolutionary synthesis, millennium synthesis or the neo-Darwinian synthesis) is a 20th-century synthesis of ideas from several fields of biology that provides an account of evolution which is widely accepted as the current paradigm in evolutionary biology, and reflects the consensus about how evolution works.

J. B .S. Haldane

The 19th Century ideas of natural selection by Charles Darwin and Mendelian genetics by Gregor Mendel were united by Ronald Fisher, one of the three founders of population genetics, along with J. B. S. Haldane and Sewall Wright, between 1918 and 1932.

Ernst Mayr

The modern synthesis solved difficulties and confusions caused by the specialisation and poor communication between biologists in the early years of the 20th century. At its heart was the question of whether Mendelian genetics could be reconciled with gradual evolution by means of natural selection. A second issue was whether the broad-scale changes of macroevolution seen by palaeontologists could be explained by changes seen in the microevolution of local populations .

The synthesis included evidence from biologists, trained in genetics, who studied populations in the field and in the laboratory. These studies were crucial to evolutionary theory. The synthesis drew together ideas from several branches of biology which had become separated, particularly genetics, cytology, systematics, botany, morphology, ecology and paleontology.

Julian Huxley invented the term in his 1942 book, *Evolution: The Modern Synthesis*. Major figures in the modern synthesis include, Theodosius Dobzhansky, E. B. Ford, Ernst Mayr, Bernhard Rensch, Sergei Chetverikov, George Gaylord Simpson, and G. Ledyard Stebbins.

Summary of the Modern Synthesis

The modern synthesis bridged the gap between the work of experimental geneticists and naturalists, and paleontologists. It states that:

1. All evolutionary phenomena can be explained in a way consistent with known genetic mechanisms and the observational evidence of naturalists.

2. Evolution is gradual: small genetic changes regulated by natural selection accumulate over long periods. Discontinuities amongst species (or other taxa) are explained as originating gradually through geographical separation and extinction. This theory contrasts with the saltation theory of William Bateson (1894).

3. Natural selection is by far the main mechanism of change; even slight advantages are important when continued. The object of selection is the phenotype in its surrounding environment.

4. The role of genetic drift is equivocal. Though strongly supported initially by Dobzhansky, it was downgraded later as results from ecological genetics were obtained.

5. Thinking in terms of populations, rather than individuals, is primary: the genetic diversity existing in natural populations is a key factor in evolution. The strength of natural selection in the wild is greater than previously expected; the effect of ecological factors such as niche occupation and the significance of barriers to gene flow are all important.

6. In palaeontology, the ability to explain historical observations by extrapolation from microevolution to macroevolution is proposed. Historical contingency means explanations at different levels may exist. Gradualism does not mean constant rate of change.

The idea that speciation occurs after populations are reproductively isolated has been much debated. In plants, polyploidy must be included in any view of speciation. Formulations such as 'evolution consists primarily of changes in the frequencies of alleles between one generation and another' were proposed rather later. The traditional view is that evolutionary developmental biology (evo-devo) played little part in the synthesis, but an account of Gavin de Beer's work by Stephen J. Gould suggests he may be an exception.

Developments Leading up to the Synthesis

1859–1899

Charles Darwin's On the Origin of Species (1859) was successful in convincing most biologists that evolution had occurred, but was less successful in convincing them that natural selection was its primary mechanism. In the 19th and early 20th centuries, variations of Lamarckism, orthogenesis ('progressive' evolution), and saltationism (evolution by jumps) were discussed as alternatives. Also, Darwin did not offer a precise explanation of how new species arise. As part of the disagreement about whether natural selection alone was sufficient to explain speciation, George Romanes coined the term neo-Darwinism to refer to the version of evolution advocated by Alfred Russel Wallace and August Weismann with its heavy dependence on natural selection. Weismann and Wallace rejected the Lamarckian idea of inheritance of acquired characteristics, something that Darwin had not ruled out.

Weismann's idea was that the relationship between the hereditary material, which he called the germ plasm (German, Keimplasma), and the rest of the body (the soma) was a one-way relationship: the germ-plasm formed the body, but the body did not influence the germ-plasm, except indirectly in its participation in a population subject to natural selection. Weismann was translated into English, and though he was influential, it took many years for the full significance of his work to be appreciated. Later, after the completion of the modern synthesis, the term neo-Darwinism came to be associated with its core concept: evolution, driven by natural selection acting on variation produced by genetic mutation, and genetic recombination (chromosomal crossovers).

1900–1915

Gregor Mendel's work was re-discovered by Hugo de Vries and Carl Correns in 1900. News of this reached William Bateson in England, who reported on the paper during a presentation to the Royal Horticultural Society in May 1900. It showed that the contributions of each parent retained their integrity rather than blending with the contribution of the other parent. This reinforced a division of thought, which was already present in the 1890s. The two schools were:

- Saltationism (large mutations or jumps), favored by early Mendelians who viewed hard inheritance as incompatible with natural selection

- Biometric school: led by Karl Pearson and Walter Weldon, argued vigorously against it, saying that empirical evidence indicated that variation was continuous in most organisms, not discrete as Mendelism predicted.

The relevance of Mendelism to evolution was unclear and hotly debated, especially by Bateson, who opposed the biometric ideas of his former teacher Weldon. Many scientists believed the two theories substantially contradicted each other. This debate between the biometricians and the Mendelians continued for some 20 years and was only solved by the development of population genetics.

Thomas Hunt Morgan began his career in genetics as a saltationist, and started out trying to demonstrate that mutations could produce new species in fruit flies. However, the experimental work at his lab with the common fruit fly, *Drosophila melanogaster*, which helped establish the link between Mendelian genetics and the chromosomal theory of inheritance, demonstrated that rather than creating new species in a single step, mutations increased the genetic variation in the population.

The Foundation of Population Genetics

The first step towards the synthesis was the development of population genetics. R. A. Fisher, J. B. S. Haldane, and Sewall Wright provided critical contributions. In 1918, Fisher produced the paper "The Correlation between Relatives on the Supposition of Mendelian Inheritance," which showed how the continuous variation measured by the biometricians could be the result of the action of many discrete genetic loci. In this and subsequent papers culminating in his 1930 book *The Genetical Theory of Natural Selection*, Fisher was able to show how Mendelian genetics was, contrary to the thinking of many early geneticists, completely consistent with the idea of evolution driven by natural selection. During the 1920s, a series of papers by Haldane applied mathematical analysis to real-world examples of natural selection such as the evolution of industrial melanism in peppered moths. Haldane established that natural selection could work in the real world at a faster rate than even Fisher had assumed.

Sewall Wright focused on combinations of genes that interacted as complexes, and the effects of inbreeding on small relatively isolated populations, which could exhibit genetic drift. In a 1932 paper, he introduced the concept of an adaptive landscape in which phenomena such as cross breeding and genetic drift in small populations could push them away from adaptive peaks, which would in turn allow natural selection to push them towards new adaptive peaks. Wright's model would appeal to field naturalists such as Theodosius Dobzhansky and Ernst Mayr who were becoming aware of the importance of geographical isolation in real world populations. The work of Fisher, Haldane and Wright founded the discipline of population genetics. This is the precursor of the modern synthesis, which is an even broader coalition of ideas. One limitation of the modern synthesis version of population genetics is that it treats one gene locus at a time, neglecting genetic linkage and resulting linkage disequilibrium between loci.

The Modern Synthesis

Theodosius Dobzhansky, an emigrant from the Soviet Union to the United States, who had been a postdoctoral worker in Morgan's fruit fly lab, was one of the first to apply genetics to natural populations. He worked mostly with *Drosophila pseudoobscura*. He says pointedly: "Russia has a variety of climates from the Arctic to sub-tropical... Exclusively laboratory workers who neither possess nor wish to have any knowledge of living beings in nature were and are in a minority." Not surprisingly, there were other Russian geneticists with similar ideas, though for some time their work was known to only a few in the West. His 1937 work *Genetics and the Origin of Species* was a key step in bridging the gap between population geneticists and field naturalists. It presented the conclusions reached by Fisher, Haldane, and especially Wright in their highly mathematical papers in a form that was easily accessible to others. It also emphasized that real world populations had far more genetic variability than the early population geneticists had assumed in their models, and that genetically distinct sub-populations were important. Dobzhansky argued that natural selection worked to maintain genetic diversity as well as driving change. Dobzhansky had been influenced by his exposure in the 1920s to the work of a Russian geneticist Sergei Chetverikov who had looked at the role of recessive genes in maintaining a reservoir of genetic variability in a population before his work was shut down by the rise of Lysenkoism in the Soviet Union.

E. B. Ford's work complemented that of Dobzhansky. It was as a result of Ford's work, as well as his own, that Dobzhansky changed the emphasis in the third edition of his famous text from drift to selection. Ford was an experimental naturalist who wanted to test natural selection in nature. He virtually invented the field of research known as ecological genetics. His work on natural selection in wild populations of butterflies and moths was the first to show that predictions made by R. A. Fisher were correct. He was the first to describe and define genetic polymorphism, and to predict that human blood group polymorphisms might be maintained in the population by providing some protection against disease.

Ernst Mayr's key contribution to the synthesis was *Systematics and the Origin of Species*, published in 1942. Mayr emphasized the importance of allopatric speciation, where geographically isolated sub-populations diverge so far that reproductive isolation occurs. He was skeptical of the reality of sympatric speciation believing that geographical isolation was a prerequisite for building up intrinsic (reproductive) isolating mechanisms. Mayr also introduced the biological species concept that defined a species as a group of interbreeding or potentially interbreeding populations that were reproductively isolated from all other populations. Before he left Germany for the United States in 1930, Mayr had been influenced by the work of German biologist Bernhard Rensch. In the 1920s Rensch, who like Mayr did field work in Indonesia, analyzed the geographic distribution of polytypic species and complexes of closely related species paying particular attention to how variations between different populations correlated with local environmental factors such as differences in climate. In 1947, Rensch published *Neuere Probleme der Abstammungslehre. Die transspezifische Evolution* (1959 English translation of 2nd edition: *Evolution Above the Species Level*). This looked at how the same evolutionary mechanisms involved in speciation might be extended to explain the origins of the differences between the higher level taxa. His writings contributed to the rapid acceptance of the synthesis in Germany.

George Gaylord Simpson was responsible for showing that the modern synthesis was compatible with paleontology in his book *Tempo and Mode in Evolution* published in 1944. Simpson's work was crucial because so many paleontologists had disagreed, in some cases vigorously, with the idea that natural selection was the main mechanism of evolution. It showed that the trends of linear progression (in for example the evolution of the horse) that earlier paleontologists had used as support for neo-Lamarckism and orthogenesis did not hold up under careful examination. Instead the fossil record was consistent with the irregular, branching, and non-directional pattern predicted by the modern synthesis.

The botanist G. Ledyard Stebbins was another major contributor to the synthesis. His major work, *Variation and Evolution in Plants*, was published in 1950. It extended the synthesis to encompass botany including the important effects of hybridization and polyploidy in some kinds of plants.

Further Advances

The modern evolutionary synthesis continued to be developed and refined after the initial establishment in the 1930s and 1940s. The work of W. D. Hamilton, George C. Williams, John Maynard Smith and others led to the development of a gene-centered view of evolution in the 1960s. The synthesis as it exists now has extended the scope of the Darwinian idea of natural selection to include subsequent scientific discoveries and concepts unknown to Darwin, such as DNA and genetics, which allow rigorous, in many cases mathematical, analyses of phenomena such as kin selection, altruism, and speciation.

John Maynard Smith

In *The Selfish Gene* (1976), author Richard Dawkins argues the gene is the only true unit of selection. (Dawkins also attempts to apply evolutionary theory to non-biological entities, such as cultural memes, imagined to be subject to selective forces analogous to those affecting biological entities.)

Richard Dawkins

Others, such as Stephen Jay Gould, reject the notion that genetic entities are subject to anything other than genetic or chemical forces, (as well as the idea evolution acts on "populations" per se), reasserting the centrality of the individual organism as the true unit of selection, whose specific phenotype is directly subject to evolutionary pressures.

In 1972, the notion of gradualism in evolution was challenged by a theory of punctuated equilibrium put forward by Gould and Niles Eldredge, proposing evolutionary changes could occur in relatively rapid spurts, when selective pressures were heightened, punctuating long periods of morphological stability, as well-adapted organisms coped successfully in their respective environments.

Discovery in the 1980s of Hox genes and regulators conserved across multiple phyletic divisions began the process of addressing basic theoretical problems relating to gradualism, incremental change, and sources of novelty in evolution. Suddenly, evolutionary theorists could answer the charge that spontaneous random mutations should result overwhelmingly in deleterious changes to a fragile, monolithic genome: Mutations in homeobox regulation could safely—yet dramatically—alter morphology at a high level, without damaging coding for specific organs or tissues.

This, in turn, provided the means to model hypothetical genomic changes expressed in the phenotypes of long-extinct species, like the Late Devonian "fish with hands," *Tiktaalik*, discovered in 2004.

As these discoveries suggest, the synthesis continues to undergo regular review, drawing on insights offered by both new biotechnologies and new paleontological discoveries.

After the Synthesis

There are a number of discoveries in earth sciences and biology which have arisen since the synthesis. Listed here are some of those topics which are relevant to the evolutionary synthesis, and which seem soundly based.

Understanding of Earth History

The Earth is the stage on which the evolutionary play is performed. Darwin studied evolution in the context of Charles Lyell's geology, but our present understanding of Earth history includes some critical advances made during the last half-century.

- The age of the Earth has been revised upwards. It is now estimated to be approximately 4.54 billion years old, nearly one-third of the age of the universe. The Phanerozoic (current eon) only occupies the last one-ninth of this period of time.

- The triumph of Alfred Wegener's idea of continental drift came around 1960. The key principle of plate tectonics is that the lithosphere exists as separate and distinct tectonic plates, which ride on the fluid-like (visco-elastic solid) asthenosphere. This discovery provides a unifying theory for geology, linking phenomena such as volcanoes, earthquakes, orogeny, and providing data for many paleogeographical questions. One major question is still unclear: when did plate tectonics begin?

- Understanding of the evolution of the atmosphere of Earth has progressed. The substitution of oxygen for carbon dioxide (CO_2) in the atmosphere, which occurred in the Proterozoic, caused probably by cyanobacteria in the form of stromatolites, caused changes leading to the evolution of aerobic organisms.

- The identification of the first generally accepted fossils of microbial life was made by geologists. These rocks have been dated as about 3.465 billion years ago. Charles Doolittle Walcott was the first geologist to identify Precambrian fossil bacteria from microscopic examination of thin rock slices. He also thought stromatolites were organic in origin. His ideas were not accepted at the time, but may now be appreciated as great discoveries.

- Information about paleoclimates is increasingly available, and being used in paleontology. One example: the discovery of massive ice ages in the Proterozoic, following the great reduction of CO_2 in the atmosphere. These ice ages were immensely long, and led to a crash in microflora.

- Catastrophism and extinction events. A partial reintegration of catastrophism has occurred, and the importance of mass extinctions in large-scale evolution is now apparent. Extinction events disturb relationships between many forms of life and may remove dominant forms and release a flow of adaptive radiation amongst groups that remain. Causes include meteorite strikes (Cretaceous–Paleogene extinction event; Late Devonian extinction); flood basalt provinces (Deccan Traps at Cretaceous–Paleogene boundary; Siberian Traps at P–Tr); and other less dramatic processes.

Conclusion: Our present knowledge of earth history strongly suggests that large-scale geophysical events influenced macroevolution and megaevolution. These terms refer to evolution above the species level, including such events as mass extinctions, adaptive radiation, and the major transitions in evolution.

Symbiotic Origin of Eukaryotic Cell Structures

Once symbiosis was discovered in lichen and in plant roots (rhizobia in root nodules) in the 19th century, the idea arose that the process might have occurred more widely, and might be important in evolution. Heinrich Anton de Bary invented the concept of *symbiosis*; several Russian biologists promoted the idea; Edmund Beecher Wilson mentioned it in his book *The Cell in Development and Heredity* (1925); as did Ivan Wallin in his *Symbionticism and the Origin of Species* (1927); and there was a brief mention by Julian Huxley in 1930; all in vain because sufficient evidence was lacking. Symbiosis as a major evolutionary force was not discussed at all in the evolutionary synthesis.

The role of symbiosis in cell evolution was revived partly by Joshua Lederberg, and finally brought to light by Lynn Margulis in a series of papers and books. Some organelles are recognized as being of microbial origin: mitochondria and chloroplasts definitely, cilia, flagella and centrioles possibly, and perhaps the nuclear membrane and much of the chromosome structure as well. What is now clear is that the evolution of eukaryote cells is either caused by, or at least profoundly influenced by, symbiosis with bacterial and archaean cells in the Proterozoic.

The origin of the eukaryote cell by symbiosis in several stages was not part of the evolutionary synthesis. It is, at least on first sight, an example of megaevolution by big jumps. However, what symbiosis provided was a copious supply of heritable variation from microorganisms, which was fine-tuned over a long period to produce the cell structure we see today. This part of the process is consistent with evolution by natural selection.

Trees of Life

The ability to analyse sequence in macromolecules (protein, DNA, RNA) provides evidence of descent, and permits us to work out genealogical trees covering the whole of life, since now there are data on every major group of living organisms. This project, begun in a tentative way in the 1960s, has become a search for the universal tree or the universal ancestor, a phrase of Carl Woese. The tree that results has some unusual features, especially in its roots. There are two domains of prokaryotes: bacteria and archaea, both of which contributed genetic material to the eukaryotes, mainly by means of symbiosis. Also, since bacteria can pass genetic material to other bacteria, their relationships look more like a web than a tree. Once eukaryotes were established, their sexual reproduction produced the traditional branching tree-like pattern, the only diagram Darwin published in *On the Origin of Species*. The last universal ancestor (LUA) would be a prokaryotic cell before the split between the bacteria and archaea. LUA is defined as most recent organism from which all organisms now living on Earth descend (some 3.5 to 3.8 billion years ago, in the Archean eon).

This technique may be used to clarify relationships within any group of related organisms. It is now a standard procedure, and examples are published regularly. April 2008 sees the publication of a tree covering all the animal phyla, derived from sequences from 150 genes in 77 taxa.

Early attempts to identify relationships between major groups were made in the 19th century by Ernst Haeckel, and by comparative anatomists such as Thomas Henry Huxley and E. Ray Lankester. Enthusiasm waned: it was often difficult to find evidence to adjudicate between different opinions. Perhaps for that reason, the evolutionary synthesis paid surprisingly little attention to this activity. It is certainly a lively field of research today.

Evolutionary Developmental Biology

What once was called embryology played a modest role in the evolutionary synthesis, mostly about evolution by changes in developmental timing (allometry and heterochrony). Man himself was, according to Louis Bolk, a typical case of evolution by retention of juvenile characteristics (neoteny). He listed many characters where "Man, in his bodily development, is a primate foetus that has become sexually mature." Unfortunately, his interpretation of these ideas was non-Darwinian, but his list of characters is both interesting and convincing.

Evolutionary developmental biology springs from clear proof that development is closely controlled by special genetic systems, and the hope that comparison of these systems will tell us much about the evolutionary history of different groups. In a series of experiments with the fruit fly *Drosophila*, Edward B. Lewis was able to identify a complex of genes whose proteins bind to the cis-regulatory regions of target genes. The latter then activate or repress systems of cellular processes that accomplish the final development of the organism. Furthermore, the sequence of these control genes show *co-linearity*: the order of the loci in the chromosome parallels the order in which the loci are expressed along the anterior-posterior axis of the body. Not only that, but this cluster of master control genes programs the development of all higher organisms. Each of the genes contains a homeobox, a remarkably conserved DNA sequence. This suggests the complex itself arose by gene duplication. In his 1995 Nobel lecture, Lewis said "Ultimately, comparisons of the HOM-C throughout the animal kingdom should provide a picture of how the organisms, as well as the genes of the HOM-C, have evolved."

The term *deep homology* was coined to describe the common origin of genetic regulatory apparatus used to build morphologically and phylogenetically disparate animal features. It applies when a complex genetic regulatory system is inherited from a common ancestor, as it is in the evolution of vertebrate and invertebrate eyes. The phenomenon is implicated in many cases of parallel evolution.

A great deal of evolution may take place by changes in the control of development. This may be relevant to punctuated equilibrium theory, for in development a few changes to the control system could make a significant difference to the adult organism. An example is the giant panda, whose place in the Carnivora was long uncertain. Apparently, the giant panda's evolution required the change of only a few genetic messages (5 or 6 perhaps), yet the phenotypic and lifestyle change from a standard bear is considerable. The transition could therefore be effected relatively swiftly.

Fossil Discoveries

Since the late 20th century, there have been excavations in parts of the world which had scarcely been investigated before. Also, there is fresh appreciation of fossils discovered in the 19th century, but then denied or deprecated: the classic example is the Ediacaran biota from the immediate Precambrian, after the Cryogenian period. These soft-bodied fossils are the first record of multicellular life. The interpretation of this fauna is still in flux.

Many outstanding discoveries have been made, and some of these have implications for evolutionary theory. The discovery of feathered dinosaurs and early birds from the Early Cretaceous of Liaoning Province in northeast China have convinced most paleontologists that birds did evolve from coelurosaurian theropod dinosaurs. Less well known, but perhaps of equal evolutionary significance, are the studies on early insect flight, on stem tetrapods from the Late Devonian, and the early stages of whale evolution.

Recent work has shed light on the evolution of flatfish (order Pleuronectiformes), such as plaice, sole, turbot and halibut. Flatfish are interesting because they are one of the few vertebrate groups with external asymmetry. Their young are perfectly symmetrical, but the head is remodelled during a metamorphosis, which entails the migration of one eye to the other side, close to the other eye. Some species have both eyes on the left (turbot), some on the right (halibut, sole); all living and fossil flatfish to date show an 'eyed' side and a 'blind' side. The lack of an intermediate condition in living and fossil flatfish species had led to debate about the origin of such a striking adaptation. The case was considered by Jean-Baptiste Lamarck, who thought flatfish precursors would have lived in shallow water for a long period, and by Charles Darwin, who predicted a gradual migration of the eye, mirroring the metamorphosis of the living forms. Darwin's long-time critic St. George Mivart thought that the intermediate stages could have no selective value, and in the 6th edition of *On the Origin of Species*, Darwin made a concession to the possibility of acquired traits. Many years later the geneticist Richard Goldschmidt put the case forward as an example of evolution by saltation, bypassing intermediate forms.

A recent examination of two fossil species from the Eocene has provided the first clear picture of flatfish evolution. The discovery of stem flatfish with incomplete orbital migration refutes Goldschmidt's ideas, and demonstrates that "the assembly of the flatfish bodyplan occurred in a gradual, stepwise fashion." There are no grounds for thinking that incomplete orbital migration was maladaptive, because stem forms with this condition ranged over two geological stages, and are found in localities which also yield flatfish with the full cranial asymmetry. The evolution of flatfish falls squarely within the evolutionary synthesis.

Horizontal Gene Transfer

Horizontal gene transfer (HGT) (or *lateral gene transfer*) is any process in which an organism gets genetic material from another organism without being the offspring of that organism.

Most thinking in genetics has focused on vertical transfer, but there is a growing awareness that horizontal gene transfer is a significant phenomenon. Amongst single-celled organisms it may be the dominant form of genetic transfer. Artificial horizontal gene transfer is a form of genetic engineering.

Aaron O. Richardson and Jeffrey D. Palmer state: "Horizontal gene transfer (HGT) has played a major role in bacterial evolution and is fairly common in certain unicellular eukaryotes. However, the prevalence and importance of HGT in the evolution of multicellular eukaryotes remain unclear."

The bacterial means of HGT are:

- Transformation, the genetic alteration of a cell resulting from the introduction, uptake and expression of foreign genetic material (DNA or RNA).

- Transduction, the process in which bacterial DNA is moved from one bacterium to another by a bacterial virus (a bacteriophage, or 'phage').

- Bacterial conjugation, a process in which a bacterial cell transfers genetic material to another cell by cell-to-cell contact.

- Gene transfer agent (GTA) is a virus-like element which contains random pieces of the host chromosome. They are found in most members of the Alphaproteobacteria order Rhodobacterales. They are encoded by the host genome. GTAs transfer DNA so frequently that they may have an important role in evolution. A 2010 report found that genes for antibiotic resistance could be transferred by engineering GTAs in the laboratory.

Some examples of HGT in Metazoa are now known. Genes in bdelloid rotifers have been found which appear to have originated in bacteria, fungi, and plants. This suggests they arrived by horizontal gene transfer. The capture and use of exogenous (~foreign) genes may represent an important force in bdelloid evolution. The team led by Matthew S. Meselson at Harvard University has also shown that, despite the lack of sexual reproduction, bdelloid rotifers do engage in genetic (DNA) transfer within a species or clade. The method used is not known at present.

Adaptation

In biology, an *adaptation*, also called an *adaptive trait*, is a trait with a current functional role in the life of an organism that is maintained and evolved by means of natural selection. Adaptation refers to both the current state of being adapted and to the dynamic evolutionary process that leads to the adaptation. Adaptations enhance the fitness and survival of individuals. Organisms face a succession of environmental challenges as they grow and develop and are equipped with an adaptive plasticity as the phenotype of traits develop in response to the imposed conditions. The developmental norm of reaction for any given trait is essential to the correction of adaptation as it affords a kind of biological insurance or resilience to varying environments.

General Principles

Adaptation is, first of all, a *process*, rather than a part of a body. An internal parasite (such as a liver fluke) can illustrate the distinction: such a parasite may have a very simple bodily structure, but nevertheless the organism is highly adapted to its specific environment. From this we see that adaptation is not just a matter of visible traits: in such parasites critical adaptations take place in the life cycle, which is often quite complex. However, as a practical term, adaptation is often used for the *product*: those features of a species which result from the process. Many aspects of an animal or plant can be correctly called adaptations, though there are always some features whose function is in doubt. By using the term *adaptation* for the evolutionary *process*, and *adaptive trait* for the bodily part or function (the product), the two senses of the word may be distinguished.

Adaptation is one of the two main processes that explain the diverse species we see in biology, such as the different species of Darwin's finches. The other is speciation (species-splitting or cladogenesis), caused by geographical isolation or some other mechanism. A favorite example used today to study the interplay of adaptation and speciation is the evolution of cichlid fish in African lakes, where the question of reproductive isolation is much more complex.

Adaptation is not always a simple matter, where the ideal phenotype evolves for a given external environment. An organism must be viable at all stages of its development and at all stages of its evolution. This places *constraints* on the evolution of development, behavior and structure of organisms. The main constraint, over which there has been much debate, is the requirement that each genetic and phenotypic change during evolution should be relatively small, because developmental systems are so complex and interlinked. However, it is not clear what "relatively small" should mean, for example polyploidy in plants is a reasonably common large genetic change. The origin of eukaryotic symbiosis is a more dramatic example.

All adaptations help organisms survive in their ecological niches. These adaptive traits may be structural, behavioral or physiological. Structural adaptations are physical features of an organism (shape, body covering, armament; and also the internal organization). Behavioral adaptations are composed of inherited behavior chains and/or the ability to learn: behaviors may be inherited in detail (instincts), or a capacity for learning may be inherited. Examples: searching for food, mating, vocalizations. Physiological adaptations permit the organism to perform special functions (for instance, making venom, secreting slime, phototropism); but also more general functions such as growth and development, temperature regulation, ionic balance and other aspects of homeostasis. Adaptation, then, affects all aspects of the life of an organism.

Definitions

The following definitions are mainly due to Theodosius Dobzhansky.

1. *Adaptation* is the evolutionary process whereby an organism becomes better able to live in its habitat or habitats.

2. *Adaptedness* is the state of being adapted: the degree to which an organism is able to live and reproduce in a given set of habitats.

3. An *adaptive trait* is an aspect of the developmental pattern of the organism which enables or enhances the probability of that organism surviving and reproducing.

Adaptedness and Fitness

From the above definitions, it is clear that there is a relationship between adaptedness and *fitness* (a key population genetics concept). Differences in fitness between genotypes predict the rate of evolution by natural selection. Natural selection changes the relative frequencies of alternative phenotypes, insofar as they are heritable. Although the two are connected, the one does not imply the other: a phenotype with high adaptedness may not have high fitness. Dobzhansky mentioned the example of the Californian redwood, which is highly adapted, but a relict species in danger of extinction. Elliott Sober commented that adaptation was a retrospective concept since it implied something about the history of a trait, whereas fitness predicts a trait's future.

1. Relative fitness. The average contribution to the next generation by a genotype or a class of genotypes, relative to the contributions of other genotypes in the population. This is also known as *Darwinian fitness*, *selection coefficient*, and other terms.

2. Absolute fitness. The absolute contribution to the next generation by a genotype or a class of genotypes. Also known as the Malthusian parameter when applied to the population as a whole.

3. Adaptedness. The extent to which a phenotype fits its local ecological niche. This can sometimes be tested through a reciprocal transplant.

Brief History

Adaptation is a fact of life that has been accepted by many of the great thinkers who have tackled the world of living organisms. It is their explanations of how adaptation arises that separates these thinkers. A few of the most significant ideas:

*   Empedocles did not believe that adaptation required a final cause (~ purpose), but "came about naturally, since such things survived." Aristotle, however, did believe in final causes.

Jean-Baptiste Lamarck (1744–1829)

*   In natural theology, adaptation was interpreted as the work of a deity, even as evidence for the existence of God. William Paley believed that organisms were perfectly adapted to the

lives they lead, an argument that shadowed Gottfried Wilhelm Leibniz, who had argued that God had brought about "the best of all possible worlds." Voltaire's Dr. Pangloss is a parody of this optimistic idea, and David Hume also argued against design. The *Bridgewater Treatises* are a product of natural theology, though some of the authors managed to present their work in a fairly neutral manner. The series was lampooned by Robert Knox, who held quasi-evolutionary views, as the *Bilgewater Treatises*. Charles Darwin broke with the tradition by emphasising the flaws and limitations which occurred in the animal and plant worlds.

- Lamarckism is a proto-evolutionary hypothesis of the inheritance of acquired characteristics, whose main purpose is to explain adaptations by natural means. Jean-Baptiste Lamarck proposed a tendency for organisms to become more complex, moving up a ladder of progress, plus "the influence of circumstances," usually expressed as *use and disuse*. His evolutionary ideas, and those of Étienne Geoffroy Saint-Hilaire, fail because they cannot be reconciled with heredity. This was known even before Gregor Mendel by medical men interested in human races (William Charles Wells, William Lawrence), and especially by August Weismann.

Many other students of natural history, such as Buffon, accepted adaptation, and some also accepted evolution, without voicing their opinions as to the mechanism. This illustrates the real merit of Darwin and Alfred Russel Wallace, and secondary figures such as Henry Walter Bates, for putting forward a mechanism whose significance had only been glimpsed previously. A century later, experimental field studies and breeding experiments by people such as E. B. Ford and Dobzhansky produced evidence that natural selection was not only the 'engine' behind adaptation, but was a much stronger force than had previously been thought.

Types of Adaptations

Adaptation is the heart and soul of evolution.

—*Niles Eldredge, Reinventing Darwin: The Great Debate at the High Table of Evolutionary Theory*

Changes in Habitat

Before Charles Darwin, adaptation was seen as a fixed relationship between an organism and its habitat. It was not appreciated that as the climate changed, so did the habitat; and as the habitat changed, so did the biota. Also, habitats are subject to changes in their biota: for example, invasions of species from other areas. The relative numbers of species in a given habitat are always changing. Change is the rule, though much depends on the speed and degree of the change.

When the habitat changes, three main things may happen to a resident population: habitat tracking, genetic change or extinction. In fact, all three things may occur in sequence. Of these three effects, only genetic change brings about adaptation.

Habitat Tracking

When a habitat changes, the most common thing to happen is that the resident population moves to another locale which suits it; this is the typical response of flying insects or oceanic organisms, who have wide (though not unlimited) opportunity for movement. This common response is called *habitat tracking*. It is one explanation put forward for the periods of apparent stasis in the fossil record (the punctuated equilibrium theory).

Genetic Change

Genetic change is what occurs in a population when natural selection acts on the genetic variability of the population; moreover, some mutations may create genetic variation that will lead to differing characteristics of offspring and hence abet adaptation. The first pathways of enzyme-based metabolism may have been parts of purine nucleotide metabolism, with previous metabolic pathways being part of the ancient RNA world. By this means, the population adapts genetically to its circumstances. Genetic changes may result in visible structures, or may adjust physiological activity in a way that suits the habitat.

It is now clear that habitats and biota do frequently change. Therefore, it follows that the process of adaptation is never finally complete. Over time, it may happen that the environment changes little, and the species comes to fit its surroundings better and better. On the other hand, it may happen that changes in the environment occur relatively rapidly, and then the species becomes less and less well adapted. Seen like this, adaptation is a genetic *tracking process*, which goes on all the time to some extent, but especially when the population cannot or does not move to another, less hostile area. Also, to a greater or lesser extent, the process affects every species in a particular ecosystem.

Leigh Van Valen thought that even in a stable environment, competing species had to constantly adapt to maintain their relative standing. This became known as the Red Queen hypothesis. One of the manifestations of the Red Queen dynamics can be seen in host-parasite interaction.

Intimate Relationships: Co-Adaptations

In coevolution, where the existence of one species is tightly over bound up with the life of another species, new or 'improved' adaptations which occur in one species are often followed by the appearance and spread of corresponding features in the other species. There are many examples of this; the idea emphasises that the life and death of living things is intimately connected, not just with the physical environment, but with the life of other species. These relationships are intrinsically dynamic, and may continue on a trajectory for millions of years, as has the relationship between flowering plants and insects (pollination).

Pollinator constancy: these honey bees selectively visit flowers from only one species, as can be seen by the colour of the pollen in their baskets:

The gut contents, wing structures, and mouthpart morphologies of fossilized beetles and flies suggest that they acted as early pollinators. The association between beetles and angiosperms during the Early Cretaceous period led to parallel radiations of angiosperms and insects into the Late Cretaceous. The evolution of nectaries in Late Cretaceous flowers signals the beginning of the mutualism between hymenopterans and angiosperms.

Mimicry

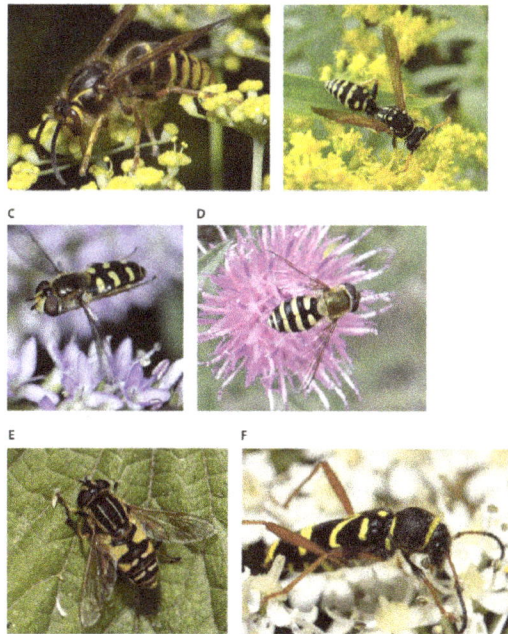

A and B show real wasps; the rest are mimics: three hoverflies and one beetle.

Bates' work on Amazonian butterflies led him to develop the first scientific account of mimicry, especially the kind of mimicry which bears his name: Batesian mimicry. This is the mimicry by a palatable species of an unpalatable or noxious species. A common example seen in temperate gardens is the hoverfly, many of which—though bearing no sting—mimic the warning colouration of hymenoptera (wasps and bees). Such mimicry does not need to be perfect to improve the survival of the palatable species.

Bates, Wallace and Fritz Müller believed that Batesian and Müllerian mimicry provided evidence for the action of natural selection, a view which is now standard amongst biologists. All aspects of this situation can be, and have been, the subject of research. Field and experimental work on these ideas continues to this day; the topic connects strongly to speciation, genetics and development.

The Basic Machinery: Internal Adaptations

There are some important adaptations to do with the overall coordination of the systems in the body. Such adaptations may have significant consequences. Examples, in vertebrates, would be temperature regulation, or improvements in brain function, or an effective immune system. An example in plants would be the development of the reproductive system in flowering plants. Such adaptations may make the clade (monophyletic group) more viable in a wide range of habitats. The acquisition of such major adaptations has often served as the spark for adaptive radiation, and huge success over long periods of time for a whole group of animals or plants.

Compromise and Conflict Between Adaptations

It is a profound truth that Nature does not know best; that genetical evolution... is a story of waste, makeshift, compromise and blunder.

— *Peter Medawar, The Future of Man*

All adaptations have a downside: horse legs are great for running on grass, but they can't scratch their backs; mammals' hair helps temperature, but offers a niche for ectoparasites; the only flying penguins do is under water. Adaptations serving different functions may be mutually destructive. Compromise and makeshift occur widely, not perfection. Selection pressures pull in different directions, and the adaptation that results is some kind of compromise.

Since the phenotype as a whole is the target of selection, it is impossible to improve simultaneously all aspects of the phenotype to the same degree.

— *Ernst Mayr, The Growth of Biological Thought: Diversity, Evolution, and Inheritance*

Consider the antlers of the Irish elk, (often supposed to be far too large; in deer antler size has an allometric relationship to body size). Obviously, antlers serve positively for defence against predators, and to score victories in the annual rut. But they are costly in terms of resource. Their size during the last glacial period presumably depended on the relative gain and loss of reproductive capacity in the population of elks during that time. Another example: camouflage to avoid detection is destroyed when vivid colors are displayed at mating time. Here the risk to life is counterbalanced by the necessity for reproduction.

Stream-dwelling salamanders, such as Caucasian salamander or Gold-striped salamander have very slender, long bodies, perfectly adapted to life at the banks of fast small rivers and mountain brooks. Elongated body protects their larvae from being washed out by current. However, elongated body increases risk of desiccation and decreases dispersal ability of the salamanders; it also negatively affects their fecundity. As a result, fire salamander, less perfectly adapted to the mountain brook habitats, is in general more successful, have a higher fecundity and broader geographic range.

An Indian peacock's trainin full display

The peacock's ornamental train (grown anew in time for each mating season) is a famous adaptation. It must reduce his maneuverability and flight, and is hugely conspicuous; also, its growth costs food resources. Darwin's explanation of its advantage was in terms of sexual selection: "This depends on the advantage which certain individuals have over other individuals of the same sex and species, in exclusive relation to reproduction." The kind of sexual selection represented by the peacock is called 'mate choice,' with an implication that the process selects the more fit over the less fit, and so has survival value. The recognition of sexual selection was for a long time in abeyance, but has been rehabilitated. In practice, the Indian peafowl (*Pavo cristatus*) is a successful species, with a large natural range in India, so the overall outcome of their mating system is quite viable.

The conflict between the size of the human foetal brain at birth, (which cannot be larger than about 400 cm^3, else it will not get through the mother's pelvis) and the size needed for an adult brain (about 1400 cm^3), means the brain of a newborn child is quite immature. The most vital things in human life (locomotion, speech) just have to wait while the brain grows and matures. That is the result of the birth compromise. Much of the problem comes from our upright bipedal stance, without which our pelvis could be shaped more suitably for birth. Neanderthals had a similar problem.

As another example, the long neck of a giraffe is a burden and a blessing. The neck of a giraffe can be up to 2 m (6 ft 7 in) in length. This neck can be used for inter-species competition or for foraging on tall trees where shorter herbivores cannot reach. However, as previously stated, there is always a trade-off. This long neck is heavy and it adds to the body mass of a giraffe, so the giraffe needs an abundance of nutrition to provide for this costly adaptation.

Shifts in Function

Adaptation and function are two aspects of one problem.

—*Julian Huxley, Evolution: The Modern Synthesis*

Pre-Adaptations

This occurs when a species or population has characteristics which (by chance) are suited for conditions which have not yet arisen. For example, the polyploid cordgrass *Spartina townsendii* is better adapted than either of its parent species to their own habitat of saline marsh and mud-flats. White Leghorn chicken are markedly more resistant to vitamin B_1 deficiency than other breeds. On a plentiful diet there is no difference, but on a restricted diet this preadaptation could be decisive.

Pre-adaptation may occur because a natural population carries a huge quantity of genetic variability. In diploid eukaryotes, this is a consequence of the system of sexual reproduction, where mutant alleles get partially shielded, for example, by the selective advantage of heterozygotes. Microorganisms, with their huge populations, also carry a great deal of genetic variability.

The first experimental evidence of the pre-adaptive nature of genetic variants in microorganisms was provided by Salvador Luria and Max Delbrück who developed Fluctuation Test, a method to show the random fluctuation of pre-existing genetic changes that conferred resistance to bacteriophage in the bacterium *Escherichia coli*.

Co-option of Existing Traits: Exaptation

The classic example is the ear ossicles of mammals, which we know from paleontological and embryological studies originated in the upper and lower jaws and the hyoid bone of their synapsid ancestors, and further back still were part of the gill arches of early fish. We owe this esoteric knowledge to the comparative anatomists, who, a century ago, were at the cutting edge of evolutionary studies. The word *exaptation* was coined to cover these shifts in function, which are surprisingly common in evolutionary history. The origin of wings from feathers that were originally used for temperature regulation is a more recent discovery.

Related Issues

Non-adaptive Traits

Some traits do not appear to be adaptive, that is, they appear to have a neutral or even deleterious effect on fitness in the current environment. Because genes have pleiotropic effects, not all traits may be functional (i.e. spandrels). Alternatively, a trait may have been adaptive at some point in an organism's evolutionary history, but a change in habitats caused what used to be an adaptation to become unnecessary or even a hindrance (maladaptations). Such adaptations are termed vestigial.

Vestigial Organs

Many organisms have vestigial organs, which are the remnants of fully functional structures in their ancestors. As a result of changes in lifestyle the organs became redundant, and are either not functional or reduced in functionality. With the loss of function goes the loss of positive selection, and the subsequent accumulation of deleterious mutations. Since any structure represents some kind of cost to the general economy of the body, an advantage may accrue from their elimination once they are not functional. Examples: wisdom teeth in humans; the loss of pigment and functional eyes in cave fauna; the loss of structure in endoparasites.

Fitness Landscapes

Sewall Wright proposed that populations occupy *adaptive peaks* on a fitness landscape. In order to evolve to another, higher peak, a population would first have to pass through a valley of maladaptive intermediate stages. A given population might be "trapped" on a peak that is not optimally adapted.

Extinction

If a population cannot move or change sufficiently to preserve its long-term viability, then obviously, it will become extinct, at least in that locale. The species may or may not survive in other locales. Species extinction occurs when the death rate over the entire species exceeds the birth rate for a long enough period for the species to disappear. It was an observation of Van Valen that groups of species tend to have a characteristic and fairly regular rate of extinction.

Coextinction

Just as we have co-adaptation, there is also coextinction. Coextinction refers to the loss of a species due to the extinction of another; for example, the extinction of parasitic insects following the loss of their hosts. Coextinction can also occur when a flowering plant loses its pollinator, or through the disruption of a food chain. Ecologist Lian Pin Koh and colleagues discuss coextinction, stating, "Species coextinction is a manifestation of the interconnectedness of organisms in complex ecosystems. . . . While coextinction may not be the most important cause of species extinctions, it is certainly an insidious one."

Flexibility, Acclimatization, Learning

Generalists, such as birds, are sometimes able to adapt to urban areas.

Flexibility deals with the relative capacity of an organism to maintain themselves in different habitats: their degree of specialization. *Acclimatization* is a term used for automatic physiological adjustments during life; *learning* is the term used for improvement in behavioral performance during life. In biology these terms are preferred, not adaptation, for changes during life which improve the performance of individuals. These adjustments are not inherited genetically by the next generation.

Adaptation, on the other hand, occurs over many generations; it is a gradual process caused by natural selection which changes the genetic make-up of a population so the collective performs better in its niche.

Flexibility

Populations differ in their phenotypic plasticity, which is the ability of an organism with a given genotype to change its phenotype in response to changes in its habitat, or to move to a different habitat.

To a greater or lesser extent, all living things can adjust to circumstances. The degree of flexibility is inherited, and varies to some extent between individuals. A highly specialized animal or plant lives only in a well-defined habitat, eats a specific type of food, and cannot survive if its needs are not met. Many herbivores are like this; extreme examples are koalas which depend on *eucalyptus*, and giant pandas which require bamboo. A generalist, on the other hand, eats a range of food, and can survive in many different conditions. Examples are humans, rats, crabs and many carnivores. The *tendency* to behave in a specialized or exploratory manner is inherited—it is an adaptation.

Rather different is *developmental flexibility:* "An animal or plant is developmentally flexible if when it is raised in or transferred to new conditions, it changes in structure so that it is better fitted to survive in the new environment," writes evolutionary biologist John Maynard Smith. Once again, there are huge differences between species, and the *capacities* to be flexible are inherited.

Acclimatization

If humans move to a higher altitude, respiration and physical exertion become a problem, but after spending time in high altitude conditions they *acclimatize* to the pressure by increasing production of Red blood cells. The *ability* to acclimatize is an adaptation, but not the acclimatization itself. Fecundity goes down, but deaths from some tropical diseases also goes down.

Over a longer period of time, some people will reproduce better at these high altitudes than others. They will contribute more heavily to later generations. Gradually the whole population becomes adapted to the new conditions. This we know takes place, because the performance of long-term communities at higher altitude is significantly better than the performance of new arrivals, even when the new arrivals have had time to make physiological adjustments.

Some kinds of acclimatization happen so rapidly that they are better called reflexes. The rapid colour changes in some flatfish, cephalopods, chameleons are examples.

Learning

Social learning is supreme for humans, and is possible for quite a few mammals and birds: of course, that does not involve genetic transmission except to the extent that the capacities are inherited. Similarly, the *capacity to learn* is an inherited adaptation, but not what is learnt; the capacity for human speech is inherited, but not the details of language.

Diversity of Genome DNAs

A large diversity of genome DNAs in a species is the basis for species' adaptation and for species' differentiation. A great number of individuals are needed for carrying the great number of different genome DNAs. According to the Misrepair-accumulation aging theory, Misrepair mechanism is important in maintaining the sufficient number of individuals in a species. Misrepair is a way of repair for increasing the surviving chance of an organism when it has severe injuries. Without Misrepairs, no individual could survive to reproduction age. Thus Misrepair mechanism is an essential mechanism for the survival of a species and for maintaining the number of individuals. Although individuals die from aging, genome DNAs are being recopied and transmitted by individuals generation by generation. In addition, the DNA Misrepairs in germ cells contribute also to the diversity of genome DNAs.

Function and Teleonomy

Adaptation raises some issues concerning how biologists use key terms such as *function*.

Function

To say something has a function is to say something about what it does for the organism, obviously. It also says something about its history: how it has come about. A heart pumps blood: that is its function. It also emits sound, which is just an ancillary side-effect. That is not its function. The heart has a history (which may be well or poorly understood), and that history is about how natural selection formed and maintained the heart as a pump. Every aspect of an organism that has a function has a history. Now, an adaptation must have a functional history: therefore we expect it must have undergone selection caused by relative survival in its habitat. It would be quite wrong to use the word adaptation about a trait which arose as a by-product.

It is widely regarded as unprofessional for a biologist to say something like "A wing is for flying," although that is their normal function. A biologist would be conscious that sometime in the remote past feathers on a small dinosaur had the function of retaining heat, and that later many wings were not used for flying (e.g. penguins, ostriches). So, the biologist would rather say that the wings on a bird or an insect usually had the *function* of aiding flight. That would carry the connotation of being an adaptation with a history of evolution by natural selection.

Teleonomy

Teleonomy is a term invented to describe the study of goal-directed functions which are not guided by the conscious forethought of man or any supernatural entity. It is contrasted with Aristotle's teleology, which has connotations of intention, purpose and foresight. Evolution is teleonomic; *adaptation hoards hindsight rather than foresight*. The following is a definition for its use in biology:

> Teleonomy: The hypothesis that adaptations arise without the existence of a prior purpose, but by chance may change the fitness of an organism.

The term may have been suggested by Colin Pittendrigh in 1958; it grew out of cybernetics and self-organising systems. Ernst Mayr, George C. Williams and Jacques Monod picked up the term and used it in evolutionary biology.

Philosophers of science have discussed the concept. Ernest Nagel analysed goal-directedness in biology; and David Hull commented on the use of teleology in biology:

> ...Haldane can be found remarking, 'Teleology is like a mistress to a biologist: he cannot live without her but he's unwilling to be seen with her in public.' Today the mistress has become a lawfully wedded wife. Biologists no longer feel obligated to apologize for their use of teleological language; they flaunt it. The only concession which they make to its disreputable past is to rename it 'teleonomy'.

Genetic Variation

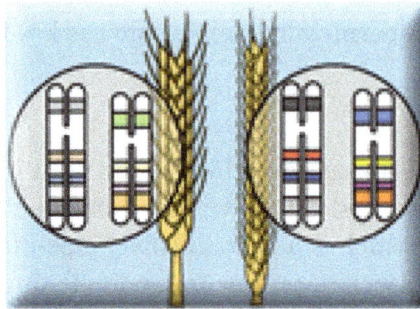

Darwin's finches or Galapagos finches.

Genetic variation is a fact that a biological system – individual and population – is different over space. It is the base of the Genetic variability of different biological systems in space.

Genetic variation is based on the variation in alleles of genes in a gene pool. It occurs both within and among populations, supported by individual carriers of the variant genes. Genetic variation is brought about by random mutation, which is a permanent change in the chemical structure of a gene.

Among Individuals within a Population

Genetic variation among individuals within a population can be identified at a variety of levels. It is possible to identify genetic variation from observations of phenotypic variation in either quantitative traits (traits that vary continuously and are coded for by many genes (e.g., leg length in dogs)) or discrete traits (traits that fall into discrete categories and are coded for by one or a few genes (e.g., white, pink, red petal color in certain flowers)).

Genetic variation can also be identified by examining variation at the level of enzymes using the process of protein electrophoresis. Polymorphic genes have more than one allele at each locus. Half of the genes that code for enzymes in insects and plants may be polymorphic, whereas polymorphisms are less common in vertebrates.

Ultimately, genetic variation is caused by variation in the order of bases in the nucleotides in genes. New technology now allows scientists to directly sequence DNA which has identified even more genetic variation than was previously detected by protein electrophoresis. Examination of DNA has shown genetic variation in both coding regions and in the non-coding intron region of genes.

Genetic variation will result in phenotypic variation if variation in the order of nucleotides in the DNA sequence results in a difference in the order of amino acids in proteins coded by that DNA sequence, and if the resultant differences in amino acid sequence influence the shape, and thus the function of the enzyme.

Between Populations

Geographic variation in genes often occurs among populations living in different locations. Geographic variation may be due to differences in selective pressures or to genetic drift.

Measurement

Genetic variation within a population is commonly measured as the percentage of gene loci that are polymorphic or the percentage of gene loci in individuals that are heterozygous.

Sources

Random mutations are the ultimate source of genetic variation. Mutations are likely to be rare and most mutations are neutral or deleterious, but in some instances the new alleles can be favored by natural selection.

A range of variability in the mussel *Donax variabilis*

Variability of walnuts

Polyploidy is an example of chromosomal mutation. Polyploidy is a condition wherein organisms have three or more sets of genetic variation (3n or more). The mutation is started off by a parent, as the parent mates the offspring now has a chance to receive that mutation trait also. Now when that mutated offspring is ready to mate they now have the chance of passing on that trait to their offspring. This process begins the first generation of mutated offspring.

Crossing over and random segregation during meiosis can result in the production of new alleles or new combinations of alleles. Furthermore, random fertilization also contributes to variation.

Variation and recombination can be facilitated by transposable and transposed genetic elements, commonly known as endogenous retroviruses, LINEs, SINEs, etc.

For a given genome of a multicellular organism, genetic variation may be acquired in somatic cells or inherited through the germline.

Forms

Genetic variation can be divided into different forms according to the size and type of genomic variation underpinning genetic change. These include:

- Small-scale sequence variation (< 1Kbp):

  o Substitution

  o Indels

- Large-scale Structural variation (>1Kbp):

  o Copy number variation:

    ▪ Copy number loss

    ▪ Copy number gain

  o Rearrangement:

    ▪ Translocation

    ▪ Inversion

    ▪ Segmental acquired uniparental disomy

- Numerical variation (whole chromosomes or genomes):

  o Polyploidy

  o Aneuploidy

Maintenance in Populations

A variety of factors maintain genetic variation in populations. Potentially harmful recessive alleles can be hidden from selection in the heterozygous individuals in populations of diploid organisms (recessive alleles are only expressed in the less common homozygous individuals). Natural selection can also maintain genetic variation in balanced polymorphisms. Balanced polymorphisms may occur when heterozygotes are favored or when selection is frequency dependent.

References

- Gould, Stephen Jay (1980). "A Quahog is a Quahog.". in: The Panda's thumb. More reflections in natural history. New York: W. W. Norton & Company. pp. 204–213. ISBN 0-393-30023-4.

- Ridley, Mark. "Speciation - What is the role of reinforcement in speciation?". Retrieved 2015-09-07. Adapted from Evolution (2004), 3rd edition (Malden, MA: Blackwell Publishing), ISBN 978-1-4051-0345-9.

- Rieger R. Michaelis A., Green M. M. (1976): Glossary of genetics and cytogenetics: Classical and molecular. Springer-Verlag, Heidelberg - New York, ISBN 3-540-07668-9; ISBN 0-387-07668-9.

- Cavali-Sforza L. L., Bodmer W. F. (1999): The genetics of human populations. Dover Publications, Inc., Mineola, New York, ISBN 0-486-40693-8.

- Rabajante, J; et al. (2016). "Host-parasite Red Queen dynamics with phase-locked rare genotypes". Science Advances. 2: e1501548. doi:10.1126/sciadv.1501548. ISSN 2375-2548.

- Van Valen, Leigh (July 1973). "A New Evolutionary Law" (PDF). Evolutionary Theory. Chicago, IL: University of Chicago. 1: 1–30. ISSN 0093-4755. Retrieved 2015-08-22.

- Wang-Michelitsch, Jicun; Michelitsch, Thomas M. (2015). "Aging as a process of accumulation of Misrepairs". arXiv:1503.07163 [q-bio.TO].

- Wang-Michelitsch, Jicun; Michelitsch, Thomas M. (2015). "Misrepair mechanism: a mechanism essential for individual adaptation, species adaptation and species evolution". arXiv:1505.03900 [q-bio.TO].

- Borenstein, Seth (13 November 2013). "Oldest fossil found: Meet your microbial mom". Excite. Yonkers, NY: Mindspark Interactive Network. Associated Press. Retrieved 2015-11-22.

- Minkel, J. R. (September 8, 2006). "Wandering Fly Gene Supports New Model of Speciation". Scientific American. Stuttgart: Georg von Holtzbrinck Publishing Group. ISSN 0036-8733. Retrieved 2015-09-11.

- Pinker, Steven (June 18, 2012). "The False Allure of Group Selection". edge.org. Edge Foundation, Inc. Retrieved 2015-09-15.

- Smith, Charles H. "Rensch, Bernhard (Carl Emmanuel) (Germany 1900-1990)". Some Biogeographers, Evolutionists and Ecologists: Chrono-Biographical Sketches. Bowling Green, KY: Western Kentucky University. Retrieved 2015-05-22.

Principles of Evolutionary Biology

Natural selection is the key mechanism of evolution while heredity is the genetic information that is passed on from the parents to the offspring. The principles of evolutionary biology are natural selection, heredity and genotype. This chapter helps in understanding the diverse topics related to evolutionary biology.

Natural Selection

Natural selection is the differential survival and reproduction of individuals due to differences in phenotype. It is a key mechanism of evolution, the change in heritable traits of a population over time. Charles Darwin popularised the term "natural selection"; he compared it with artificial selection (selective breeding).

Variation exists within all populations of organisms. This occurs partly because random mutations arise in the genome of an individual organism, and offspring can inherit such mutations. Throughout the lives of the individuals, their genomes interact with their environments to cause variations in traits. (The environment of a genome includes the molecular biology in the cell, other cells, other individuals, populations, species, as well as the abiotic environment.) Individuals with certain variants of the trait may survive and reproduce more than individuals with other, less successful, variants. Therefore, the population evolves. Factors that affect reproductive success are also important, an issue that Darwin developed in his ideas on sexual selection (now often included in natural selection) and on fecundity selection, for example.

Natural selection acts on the phenotype, or the observable characteristics of an organism, but the genetic (heritable) basis of any phenotype that gives a reproductive advantage may become more common in a population. Over time, this process can result in populations that specialise for particular ecological niches (microevolution) and may eventually result in the emergence of new species (macroevolution). In other words, natural selection is an important process (though not the only process) by which evolution takes place within a population of organisms. Natural selection can be contrasted with artificial selection, in which humans intentionally choose specific traits (although they may not always get what they want). In natural selection there is no intentional choice. In other words, artificial selection is teleological and natural selection is not teleological, though biologists often use teleological language to describe it.

Natural selection is one of the cornerstones of modern biology. The concept, published by Darwin and Alfred Russel Wallace in a joint presentation of papers in 1858, was elaborated in Darwin's influential 1859 book *On the Origin of Species*, which described natural selection as analogous to artificial selection, a process by which animals and plants with traits considered desirable by human breeders are systematically favoured for reproduction. The concept of natural selection originally developed in the absence of a valid theory of heredity; at the time of Darwin's writing, science had yet to develop modern theories of genetics. The union of traditional Darwinian evolution with subsequent discoveries in classical and molecular genetics is termed the *modern evolutionary synthesis*. Natural selection remains the primary explanation for adaptive evolution.

General Principles

Natural variation occurs among the individuals of any population of organisms. Many of these differences do not affect survival or reproduction, but some differences may improve the chances of survival and reproduction of a particular individual. A rabbit that runs faster than others may be more likely to escape from predators, and algae that are more efficient at extracting energy from sunlight will grow faster. Something that increases an organism's chances of survival will often also include its reproductive rate; however, sometimes there is a trade-off between survival and current reproduction. Ultimately, what matters is total lifetime reproduction of the organism.

Male peppered moth

Morpha *typica* and morpha *carbonaria*, morphs of the peppered moth resting on the same tree.

The peppered moth exists in both light and dark colours in the United Kingdom, but during the industrial revolution, many of the trees on which the moths rested became blackened by soot, giving the dark-coloured moths an advantage in hiding from predators. This gave dark-coloured moths a better chance of surviving to produce dark-coloured offspring, and in just fifty years from the first dark moth being caught, nearly all of the moths in industrial Manchester were dark. The balance was reversed by the effect of the Clean Air Act 1956, and the dark moths became rare again, demonstrating the influence of natural selection on peppered moth evolution.

If the traits that give these individuals a reproductive advantage are also heritable, that is, passed from parent to offspring, then there will be a slightly higher proportion of fast rabbits or efficient algae in the next generation. This is known as *differential reproduction*. Even if the reproductive advantage is very slight, over many generations any heritable advantage will become dominant in the population. In this way the natural environment of an organism "selects" for traits that confer a reproductive advantage, causing gradual changes or evolution of life. This effect was first described and named by Charles Darwin.

The concept of natural selection predates the understanding of genetics, the mechanism of heredity for all known life forms. In modern terms, selection acts on an organism's phenotype, or observable characteristics, but it is the organism's genetic make-up or genotype that is inherited. The phenotype is the result of the genotype and the environment in which the organism lives.

This is the link between natural selection and genetics, as described in the modern evolutionary synthesis. Although a complete theory of evolution also requires an account of how genetic variation arises in the first place (such as by mutation and sexual reproduction) and includes other evolutionary mechanisms (such as genetic drift and gene flow), natural selection appears to be the most important mechanism for creating complex adaptations in nature.

Nomenclature and Usage

The term *natural selection* has slightly different definitions in different contexts. It is most often defined to operate on heritable traits, because these are the traits that directly participate in evolution. However, natural selection is "blind" in the sense that changes in phenotype (physical and behavioural characteristics) can give a reproductive advantage regardless of whether or not the trait is heritable (non heritable traits can be the result of environmental factors or the life experience of the organism).

Following Darwin's primary usage the term is often used to refer to both the evolutionary consequence of blind selection and to its mechanisms. It is sometimes helpful to explicitly distinguish between selection's mechanisms and its effects; when this distinction is important, scientists define "natural selection" specifically as "those mechanisms that contribute to the selection of individuals that reproduce," without regard to whether the basis of the selection is heritable. This is sometimes referred to as "phenotypic natural selection."

Traits that cause greater reproductive success of an organism are said to be selected for, whereas those that reduce success are selected against. Selection for a trait may also result in the selection of other correlated traits that do not themselves directly influence reproductive advantage. This may occur as a result of pleiotropy or gene linkage.

Fitness

The concept of fitness is central to natural selection. In broad terms, individuals that are more "fit" have better potential for survival, as in the well-known phrase "survival of the fittest." However, as with natural selection above, the precise meaning of the term is much more subtle. Modern evolutionary theory defines fitness not by how long an organism lives, but by how successful it is at reproducing. If an organism lives half as long as others of its species, but has twice as many offspring surviving to adulthood, its genes will become more common in the adult population of the next generation.

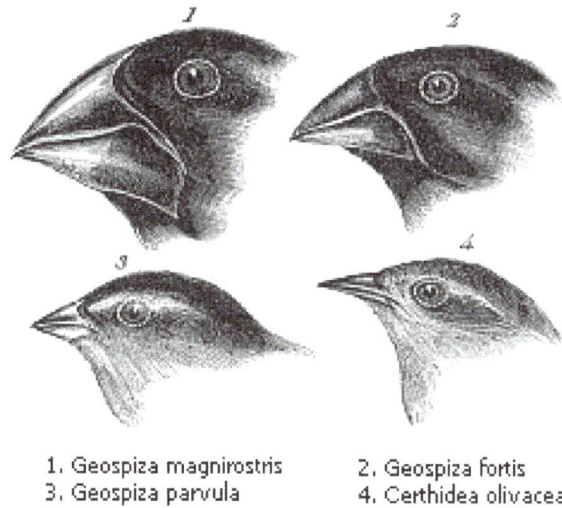

1. Geospiza magnirostris   2. Geospiza fortis
3. Geospiza parvula   4. Certhidea olivacea

Finches from Galapagos Archipelago

Charles Darwin's illustrations of beak variation in the finches of the Galápagos Islands, which hold 13 closely related species that differ most markedly in the shape of their beaks. The beak of each species is suited to its preferred food, suggesting that beak shapes evolved by natural selection.

Though natural selection acts on individuals, the effects of chance mean that fitness can only really be defined "on average" for the individuals within a population. The fitness of a particular genotype corresponds to the average effect on all individuals with that genotype. Very low-fitness genotypes cause their bearers to have few or no offspring on average; examples include many human genetic disorders like cystic fibrosis.

Since fitness is an averaged quantity, it is also possible that a favourable mutation arises in an individual that does not survive to adulthood for unrelated reasons. Fitness also depends crucially upon the environment. Conditions like sickle-cell anaemia may have low fitness in the general human population, but because the sickle cell trait confers immunity from malaria, it has high fitness value in populations that have high malaria infection rates.

Types of Selection

Natural selection can act on any heritable phenotypic trait, and selective pressure can be produced by any aspect of the environment, including sexual selection and competition with members of the same or other species. However, this does not imply that natural selection is always directional and results in adaptive evolution; natural selection often results in the maintenance of the status quo by eliminating less fit variants.

Selection can be classified according to its effect on a trait. Stabilizing selection acts to hold a trait at a stable optimum, and in the simplest case all deviations from this optimum are selectively disadvantageous. Directional selection acts during transition periods, when the current mode of the trait is sub-optimal, and alters the trait towards a single new optimum. Disruptive selection also acts during transition periods when the current mode is sub-optimal, but alters the trait in more than one direction. In particular, if the trait is quantitative and univariate then both higher and lower trait levels are favoured. Disruptive selection can be a precursor to speciation.

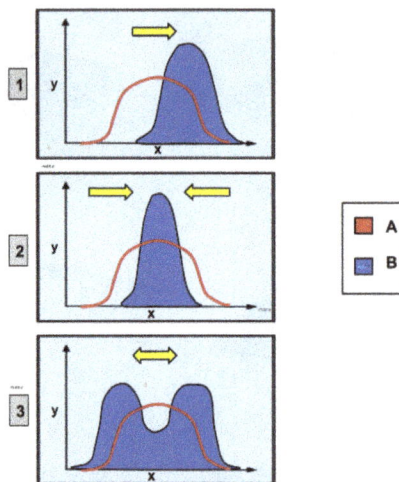

These charts depict the different types of genetic selection. On each graph, the x-axis variable is the type of phenotypic trait and the y-axis variable is the amount of organisms. Group A is the original population and Group B is the population after selection. Graph 1 shows directional selection, in which a single extreme phenotype is favored. Graph 2 depicts stabilizing selection, where the intermediate phenotype is favored over the extreme traits. Graph 3 shows disruptive selection, in which the extreme phenotypes are favored over the intermediate.

Selection can also be classified according to its effect on allele frequency. Positive selection acts to increase the frequency of an allele. Negative selection acts to decrease the frequency of an allele. Note that for a diallelic locus, positive selection on one allele perforce implies negative selection on the other allele.

Selection can also be classified according to its effect on genetic diversity. Purifying selection acts to remove genetic variation from the population (and is opposed by *de novo* mutation, which introduces new variation). Balancing selection acts to maintain genetic variation in a population (even in the absence of *de novo* mutation). Mechanisms include negative frequency-dependent selection (of which heterozygote advantage is a special case), and spatial and/or temporal fluctuations in the strength and direction of selection.

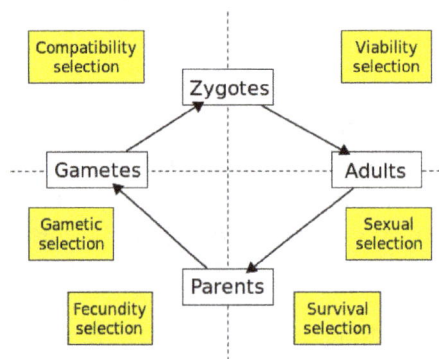

Released under public domain, http://en.wikipedia.org/wiki/User:Wykis

The life cycle of a sexually reproducing organism. Various components of natural selection are indicated for each life stage.

Selection can also be classified according to the stage of an organism's life cycle at which it acts. The use of terminology differs here. Some recognise just two types of selection: viability selection (or survival selection) which acts to improve the probability of survival of the organism, and fecundity selection (or fertility selection, or reproductive selection) which acts to improve the rate of reproduction, given successful survival. Others split the life cycle into further components of selection. Thus viability and survival selection may be defined separately and respectively as acting to improve the probability of survival before and after reproductive age is reached, while fecundity selection may be split into additional sub-components including sexual selection, *gametic selection* (acting on gamete survival) and *compatibility selection* (acting on zygote formation).

Selection can also be classified according to the level or unit of selection. *Individual selection* acts at the level of the individual, in the sense that adaptions are 'for' the benefit of the individual, and result from selection among individuals. Gene selection acts directly at the level of the gene. In many situations, this is simply a different way of describing individual selection. However, in some cases (e.g., kin selection and intragenomic conflict), gene-level selection provides a more apt explanation of the underlying process. Group selection acts at the level of groups of organisms. The mechanism assumes that groups replicate and mutate in an analogous way to genes and individuals. There is an ongoing debate over the degree to which group selection occurs in nature.

Finally, selection can be classified according to the resource being competed for. Sexual selection results from competition for mates. Sexual selection can be *intrasexual*, as in cases of competition among individuals of the same sex in a population, or *intersexual*, as in cases where one sex controls reproductive access by choosing among a population of available mates. Typically, sexual selection proceeds via fecundity selection, sometimes at the expense of viability. Ecological selection is natural selection via any other means than sexual selection. Alternatively, natural selection is sometimes defined as synonymous with ecological selection, and sexual selection is then classified as a separate mechanism to natural selection. This accords with Darwin's usage of these terms, but ignores the fact that mate competition and mate choice are natural processes.

Note that types of selection often act in concert. Thus Stabilizing selection typically proceeds via negative selection on rare alleles, leading to purifying selection, while directional selection typically proceeds via positive selection on an initially rare favoured allele.

Ecological selection covers any mechanism of selection as a result of the environment, including relatives (e.g., kin selection, competition, and infanticide).

Sexual Selection

The peacock tail in flight, a classic example of sexual selection.

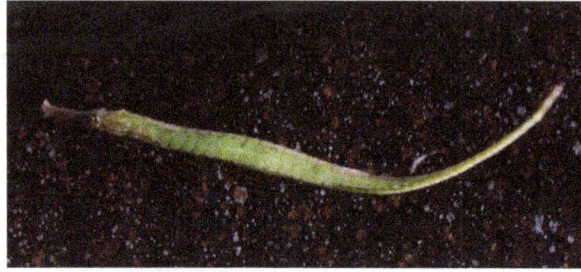

Alligator Pipefish, a syngnathid

Sexual selection refers specifically to competition for mates, which can be *intrasexual*, between individuals of the same sex, that is male–male competition, or *intersexual*, where one gender choose mates. However, some species exhibit sex-role reversed behaviour in which it is males that are most selective in mate choice; such as in some fishes of the family Syngnathidae, though likely examples have also been found in sexual selection in amphibians, sexual selection in birds, sexual selection in mammals (including sexual selection in humans) and sexual selection in scaled reptiles.

Phenotypic traits can be displayed in one sex and desired in the other sex, causing a positive feedback loop called a Fisherian runaway, for example, the extravagant plumage of some male birds. An alternate theory proposed by the same Ronald Fisher in 1930 is the sexy son hypothesis, that mothers will want promiscuous sons to give them large numbers of grandchildren and so will choose promiscuous fathers for their children. Aggression between members of the same sex is sometimes associated with very distinctive features, such as the antlers of stags, which are used in combat with other stags. More generally, intrasexual selection is often associated with sexual dimorphism, including differences in body size between males and females of a species.

Examples of Natural Selection

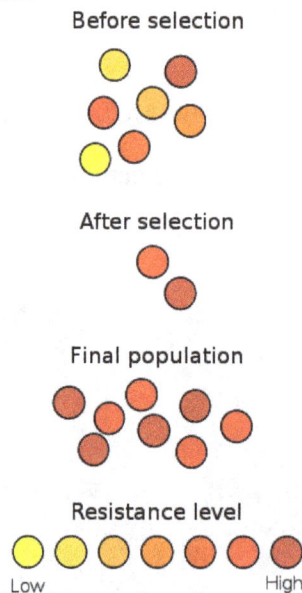

Resistance to antibiotics is increased though the survival of individuals that are immune to the effects of the antibiotic, whose offspring then inherit the resistance, creating a new population of resistant bacteria.

A well-known example of natural selection in action is the development of antibiotic resistance in microorganisms. Since the discovery of penicillin in 1928, antibiotics have been used to fight bacterial diseases. Natural populations of bacteria contain, among their vast numbers of individual members, considerable variation in their genetic material, primarily as the result of mutations. When exposed to antibiotics, most bacteria die quickly, but some may have mutations that make them slightly less susceptible. If the exposure to antibiotics is short, these individuals will survive the treatment. This selective elimination of maladapted individuals from a population is natural selection.

These surviving bacteria will then reproduce again, producing the next generation. Due to the elimination of the maladapted individuals in the past generation, this population contains more bacteria that have some resistance against the antibiotic. At the same time, new mutations occur, contributing new genetic variation to the existing genetic variation. Spontaneous mutations are very rare, and advantageous mutations are even rarer. However, populations of bacteria are large enough that a few individuals will have beneficial mutations. If a new mutation reduces their susceptibility to an antibiotic, these individuals are more likely to survive when next confronted with that antibiotic.

Given enough time and repeated exposure to the antibiotic, a population of antibiotic-resistant bacteria will emerge. This new changed population of antibiotic-resistant bacteria is optimally adapted to the context it evolved in. At the same time, it is not necessarily optimally adapted any more to the old antibiotic free environment. The end result of natural selection is two populations that are both optimally adapted to their specific environment, while both perform substandard in the other environment.

The widespread use and misuse of antibiotics has resulted in increased microbial resistance to antibiotics in clinical use, to the point that the methicillin-resistant *Staphylococcus aureus* (MRSA) has been described as a "superbug" because of the threat it poses to health and its relative invulnerability to existing drugs. Response strategies typically include the use of different, stronger antibiotics; however, new strains of MRSA have recently emerged that are resistant even to these drugs.

This is an example of what is known as an evolutionary arms race, in which bacteria continue to develop strains that are less susceptible to antibiotics, while medical researchers continue to develop new antibiotics that can kill them. A similar situation occurs with pesticide resistance in plants and insects. Arms races are not necessarily induced by man; a well-documented example involves the spread of a gene in the butterfly *Hypolimnas bolina* suppressing male-killing activity by *Wolbachia* bacteria parasites on the island of Samoa, where the spread of the gene is known to have occurred over a period of just five years

Evolution by Means of Natural Selection

A prerequisite for natural selection to result in adaptive evolution, novel traits and speciation, is the presence of heritable genetic variation that results in fitness differences. Genetic variation is the result of mutations, genetic recombinations and alterations in the karyotype (the number, shape, size and internal arrangement of the chromosomes). Any of these changes might have an effect that is highly advantageous or highly disadvantageous, but large effects are very rare. In the past, most changes in the genetic material were considered neutral or close to neutral because they occurred in noncoding DNA or resulted in a synonymous substitution. However, recent research suggests that many mutations in non-coding DNA do have slight deleterious effects. Although both mutation rates and average fitness effects of mutations are dependent on the organism, estimates from data in humans have found that a majority of mutations are slightly deleterious.

The exuberant tail of the peacock is thought to be the result of sexual selection by females. This peacock is leucistic; selection against leucism and albinism in nature is intense because they are easily spotted by predators or are unsuccessful in competition for mates.

By the definition of fitness, individuals with greater fitness are more likely to contribute offspring to the next generation, while individuals with lesser fitness are more likely to die early or fail to reproduce. As a result, alleles that on average result in greater fitness become more abundant in the next generation, while alleles that in general reduce fitness become rarer. If the selection forces remain the same for many generations, beneficial alleles become more and more abundant, until they dominate the population, while alleles with a lesser fitness disappear. In every generation, new mutations and re-combinations arise spontaneously, producing a new spectrum of phenotypes. Therefore, each new generation will be enriched by the increasing abundance of alleles that contribute to those traits that were favoured by selection, enhancing these traits over successive generations.

Some mutations occur in so-called regulatory genes. Changes in these can have large effects on the phenotype of the individual because they regulate the function of many other genes. Most, but not all, mutations in regulatory genes result in non-viable zygotes. Examples of nonlethal regulatory mutations occur in HOX genes in humans, which can result in a cervical rib or polydactyly, an increase in the number of fingers or toes. When such mutations result in a higher fitness, natural selection will favour these phenotypes and the novel trait will spread in the population.

Established traits are not immutable; traits that have high fitness in one environmental context may be much less fit if environmental conditions change. In the absence of natural selection to preserve such a trait, it will become more variable and deteriorate over time, possibly resulting in a vestigial manifestation of the trait, also called evolutionary baggage. In many circumstances, the apparently vestigial structure may retain a limited functionality, or may be co-opted for other advantageous traits in a phenomenon known as preadaptation. A famous example of a vestigial structure, the eye of the blind mole-rat, is believed to retain function in photoperiod perception.

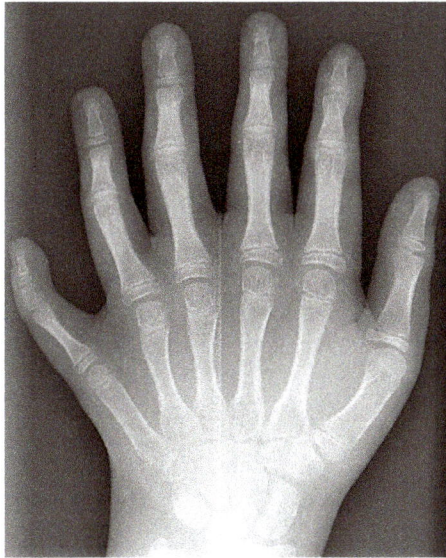

X-ray of the left hand of a ten-year-old boy with polydactyly.

Speciation

Speciation requires cessation of gene flow, which results from reproductive isolation. Over time, isolated subgroups might diverge radically to become different species, either because of differences in selection pressures on the different subgroups, or because different mutations arise spontaneously in the different populations, or because of genetic drift which is responsible for phenomena such as bottleneck effect and founder effect. A lesser-known mechanism of speciation occurs via hybridisation, well-documented in plants and occasionally observed in species-rich groups of animals such as cichlid fishes. Such mechanisms of rapid speciation can reflect a mechanism of evolutionary change known as punctuated equilibrium, which suggests that evolutionary change and in particular speciation typically happens quickly after interrupting long periods of stasis.

Genetic changes within groups result in increasing incompatibility between the genomes of the two subgroups, thus reducing gene flow between the groups. Gene flow will effectively cease when the distinctive mutations characterising each subgroup become fixed. As few as two mutations can result in speciation: if each mutation has a neutral or positive effect on fitness when they occur separately, but a negative effect when they occur together, then fixation of these genes in the respective subgroups will lead to two reproductively isolated populations. According to the biological species concept, these will be two different species.

Historical Development

Pre-Darwinian Theories

Several ancient philosophers expressed the idea that nature produces a huge variety of creatures, randomly, and that only those creatures that manage to provide for themselves and reproduce successfully survive; well-known examples include Empedocles and his intellectual successor, the

Roman poet Lucretius. Empedocles' idea that organisms arose entirely by the incidental workings of causes such as heat and cold was criticised by Aristotle in Book II of *Physics*. He posited natural teleology in its place. He believed that form was achieved for a purpose, citing the regularity of heredity in species as proof. Nevertheless, he acceded that new types of animals, monstrosities, can occur in very rare instances (*Generation of Animals*, Book IV). As quoted in Darwin's *The Origin of Species* (1872), Aristotle considered whether different forms (e.g., of teeth) might have appeared accidentally, but only the useful forms survived:

The modern theory of natural selection derives from the work of Charles Darwin in the nineteenth century.

So what hinders the different parts [of the body] from having this merely accidental relation in nature? as the teeth, for example, grow by necessity, the front ones sharp, adapted for dividing, and the grinders flat, and serviceable for masticating the food; since they were not made for the sake of this, but it was the result of accident. And in like manner as to the other parts in which there appears to exist an adaptation to an end. Wheresoever, therefore, all things together (that is all the parts of one whole) happened like as if they were made for the sake of something, these were preserved, having been appropriately constituted by an internal spontaneity, and whatsoever things were not thus constituted, perished, and still perish.

—*Aristotle, Physics, Book II, Chapter 8*

But he rejected this possibility in the next paragraph:

...Yet it is impossible that this should be the true view. For teeth and all other natural things either invariably or normally come about in a given way; but of not one of the results of chance or spontaneity is this true. We do not ascribe to chance or mere coincidence the frequency of rain in winter, but frequent rain in summer we do; nor heat in the dog-days, but only if we have it in winter. If then, it is agreed that things are either the result of coincidence or for an end, and these cannot be the result of coincidence or spontaneity, it follows that they must be for an end; and that such things are all due to nature even the champions of the theory which is before us would agree. Therefore action for an end is present in things which come to be and are by nature.

—*Aristotle, Physics, Book II, Chapter 8*

The struggle for existence was later described by Islamic writer Al-Jahiz in the 9th century.

The classical arguments were reintroduced in the 18th century by Pierre Louis Maupertuis and others, including Charles Darwin's grandfather Erasmus Darwin. While these forerunners had an influence on Darwinism, they later had little influence on the trajectory of evolutionary thought after Charles Darwin.

Until the early 19th century, the prevailing view in Western societies was that differences between individuals of a species were uninteresting departures from their Platonic idealism (or typus) of created kinds. However, the theory of uniformitarianism in geology promoted the idea that simple, weak forces could act continuously over long periods of time to produce radical changes in the Earth's landscape. The success of this theory raised awareness of the vast scale of geological time and made plausible the idea that tiny, virtually imperceptible changes in successive generations could produce consequences on the scale of differences between species.

Early 19th-century evolutionists such as Jean-Baptiste Lamarck suggested the inheritance of acquired characteristics as a mechanism for evolutionary change; adaptive traits acquired by an organism during its lifetime could be inherited by that organism's progeny, eventually causing transmutation of species. This theory has come to be known as Lamarckism and was an influence on the anti-genetic ideas of the Stalinist Soviet biologist Trofim Lysenko.

Between 1835 and 1837, zoologist Edward Blyth also contributed specifically to the area of variation, artificial selection, and how a similar process occurs in nature. In fact, Charles Darwin showed his high regards for Blyth's ideas in the first chapter on variation of *On the Origin of Species* that he wrote, "Mr. Blyth, whose opinion, from his large and varied stores of knowledge, I should value more than that of almost any one, ..."

Darwin's Theory

In 1859, Charles Darwin set out his theory of evolution by natural selection as an explanation for adaptation and speciation. He defined natural selection as the "principle by which each slight variation [of a trait], if useful, is preserved." The concept was simple but powerful: individuals best adapted to their environments are more likely to survive and reproduce. As long as there is some variation between them and that variation is heritable, there will be an inevitable selection of individuals with the most advantageous variations. If the variations are inherited, then differential reproductive success will lead to a progressive evolution of particular populations of a species, and populations that evolve to be sufficiently different eventually become different species.

Darwin's ideas were inspired by the observations that he had made on the second voyage of HMS *Beagle* (1831–1836), and by the work of a political economist, the Reverend Thomas Robert Malthus, who in *An Essay on the Principle of Population* (1798), noted that population (if unchecked) increases exponentially, whereas the food supply grows only arithmetically; thus, inevitable limitations of resources would have demographic implications, leading to a "struggle for existence." When Darwin read Malthus in 1838 he was already primed by his work as a naturalist to appreciate the "struggle for existence" in nature and it struck him that as population outgrew resources, "favourable variations would tend to be preserved, and unfavourable ones to be destroyed. The result of this would be the formation of new species."

Here is Darwin's own summary of the idea, which can be found in the fourth chapter of *On the Origin of Species*:

> If during the long course of ages and under varying conditions of life, organic beings vary at all in the several parts of their organisation, and I think this cannot be disputed; if there be, owing to the high geometrical powers of increase of each species, at some age, season, or year, a severe struggle for life, and this certainly cannot be disputed; then, considering the infinite complexity of the relations of all organic beings to each other and to their conditions of existence, causing an infinite diversity in structure, constitution, and habits, to be advantageous to them, I think it would be a most extraordinary fact if no variation ever had occurred useful to each being's own welfare, in the same way as so many variations have occurred useful to man. But if variations useful to any organic being do occur, assuredly individuals thus characterised will have the best chance of being preserved in the struggle for life; and from the strong principle of inheritance they will tend to produce offspring similarly characterised. This principle of preservation, I have called, for the sake of brevity, Natural Selection.

Once he had his theory "by which to work," Darwin was meticulous about gathering and refining evidence as his "prime hobby" before making his idea public. He was in the process of writing his "big book" to present his researches when the naturalist Alfred Russel Wallace independently conceived of the principle and described it in an essay he sent to Darwin to forward to Charles Lyell. Lyell and Joseph Dalton Hooker decided (without Wallace's knowledge) to present his essay together with unpublished writings that Darwin had sent to fellow naturalists, and *On the Tendency of Species to form Varieties; and on the Perpetuation of Varieties and Species by Natural Means of Selection* was read to the Linnean Society of London announcing co-discovery of the principle in July 1858. Darwin published a detailed account of his evidence and conclusions in *On the Origin of Species* in 1859. In the 3rd edition of 1861 Darwin acknowledged that others—a notable one being William Charles Wells in 1813, and Patrick Matthew in 1831—had proposed similar ideas, but had neither developed them nor presented them in notable scientific publications.

Darwin thought of natural selection by analogy to how farmers select crops or livestock for breeding, which he called "artificial selection"; in his early manuscripts he referred to a *Nature*, which would do the selection. At the time, other mechanisms of evolution such as evolution by genetic drift were not yet explicitly formulated, and Darwin believed that selection was likely only part of the story: "I am convinced that Natural Selection has been the main but not exclusive means of modification." In a letter to Charles Lyell in September 1860, Darwin regretted the use of the term "Natural Selection," preferring the term "Natural Preservation."

For Darwin and his contemporaries, natural selection was in essence synonymous with evolution by natural selection. After the publication of *On the Origin of Species*, educated people generally accepted that evolution had occurred in some form. However, natural selection remained controversial as a mechanism, partly because it was perceived to be too weak to explain the range of observed characteristics of living organisms, and partly because even supporters of evolution balked at its "unguided" and non-progressive nature, a response that has been characterised as the single most significant impediment to the idea's acceptance.

However, some thinkers enthusiastically embraced natural selection; after reading Darwin, Herbert Spencer introduced the term *survival of the fittest*, which became a popular summary of the theory. The fifth edition of *On the Origin of Species* published in 1869 included Spencer's phrase as an alternative to natural selection, with credit given: "But the expression often used by Mr. Herbert Spencer of the Survival of the Fittest is more accurate, and is sometimes equally convenient." Although the phrase is still often used by non-biologists, modern biologists avoid it because it is tautological if "fittest" is read to mean "functionally superior" and is applied to individuals rather than considered as an averaged quantity over populations.

Modern Evolutionary Synthesis

Natural selection relies crucially on the idea of heredity, but developed before the basic concepts of genetics. Although the Moravian monk Gregor Mendel, the father of modern genetics, was a contemporary of Darwin's, his work would lie in obscurity until the early 20th century. Only after the 20th-century integration of Darwin's theory of evolution with a complex statistical appreciation of Gregor Mendel's "re-discovered" laws of inheritance did scientists generally come to accept natural selection.

The work of Ronald Fisher (who developed the required mathematical language and wrote *The Genetical Theory of Natural Selection* (1930)), J. B. S. Haldane (who introduced the concept of the "cost" of natural selection), Sewall Wright (who elucidated the nature of selection and adaptation), Theodosius Dobzhansky (who established the idea that mutation, by creating genetic diversity, supplied the raw material for natural selection: see *Genetics and the Origin of Species* (1937)), William D. Hamilton (who conceived of kin selection), Ernst Mayr (who recognised the key importance of reproductive isolation for speciation: see *Systematics and the Origin of Species* (1942)) and many others together formed the modern evolutionary synthesis. This synthesis cemented natural selection as the foundation of evolutionary theory, where it remains today.

Genetic Basis of Natural Selection

The idea of natural selection predates the understanding of genetics. We now have a much better idea of the biology underlying heritability, which is the basis of natural selection.

Genotype and Phenotype

Natural selection acts on an organism's phenotype, or physical characteristics. Phenotype is determined by an organism's genetic make-up (genotype) and the environment in which the organism lives. Often, natural selection acts on specific traits of an individual, and the terms phenotype and genotype are used narrowly to indicate these specific traits.

When different organisms in a population possess different versions of a gene for a certain trait, each of these versions is known as an allele. It is this genetic variation that underlies phenotypic traits. A typical example is that certain combinations of genes for eye colour in humans that, for instance, give rise to the phenotype of blue eyes. (On the other hand, when all the organisms in a population share the same allele for a particular trait, and this state is stable over time, the allele is said to be *fixed* in that population.)

Some traits are governed by only a single gene, but most traits are influenced by the interactions of many genes. A variation in one of the many genes that contributes to a trait may have only a small effect on the phenotype; together, these genes can produce a continuum of possible phenotypic values.

Directionality of Selection

When some component of a trait is heritable, selection will alter the frequencies of the different alleles, or variants of the gene that produces the variants of the trait. Selection can be divided into three classes, on the basis of its effect on allele frequencies.

Directional selection occurs when a certain allele has a greater fitness than others, resulting in an increase of its frequency. This process can continue until the allele is fixed and the entire population shares the fitter phenotype. It is directional selection that is illustrated in the antibiotic resistance example above.

Far more common is Stabilizing selection (which is commonly confused with *purifying selection*), which lowers the frequency of alleles that have a deleterious effect on the phenotype – that is, produce organisms of lower fitness. This process can continue until the allele is eliminated from the population. Purifying selection results in functional genetic features, such as protein-coding genes or regulatory sequences, being conserved over time due to selective pressure against deleterious variants.

A number of forms of balancing selection exist, which do not result in fixation, but maintain an allele at intermediate frequencies in a population. This can occur in diploid species (that is, those that have homologous pairs of chromosomes) when heterozygote individuals, who have different alleles on each chromosome at a single genetic locus, have a higher fitness than homozygote individuals that have two of the same alleles. This is called heterozygote advantage or over-dominance, of which the best-known example is the malarial resistance observed in heterozygous humans who carry only one copy of the gene for sickle-cell anaemia. Maintenance of allelic variation can also occur through disruptive or diversifying selection, which favours genotypes that depart from the average in either direction (that is, the opposite of over-dominance), and can result in a bimodal distribution of trait values. Finally, balancing selection can occur through frequency-dependent selection, where the fitness of one particular phenotype depends on the distribution of other phenotypes in the population. The principles of game theory have been applied to understand the fitness distributions in these situations, particularly in the study of kin selection and the evolution of reciprocal altruism.

Selection and Genetic Variation

A portion of all genetic variation is functionally neutral in that it produces no phenotypic effect or significant difference in fitness; the hypothesis that this variation accounts for a large fraction of observed genetic diversity is known as the neutral theory of molecular evolution and was originated by Motoo Kimura. When genetic variation does not result in differences in fitness, selection cannot *directly* affect the frequency of such variation. As a result, the genetic variation at those sites will be higher than at sites where variation does influence fitness. However, after a period with no new mutation, the genetic variation at these sites will be eliminated due to genetic drift.

Mutation–Selection Balance

Natural selection results in the reduction of genetic variation through the elimination of maladapted individuals and consequently of the mutations that caused the maladaptation. At the same time, new mutations occur, resulting in a Mutation–selection balance. The exact outcome of the two processes depends both on the rate at which new mutations occur and on the strength of the natural selection, which is a function of how unfavourable the mutation proves to be. Consequently, changes in the mutation rate or the selection pressure will result in a different Mutation–selection balance.

Genetic Linkage

Genetic linkage occurs when the loci of two alleles are *linked*, or in close proximity to each other on the chromosome. During the formation of gametes, recombination of the genetic material results in reshuffling of the alleles. However, the chance that such a reshuffle occurs between two alleles depends on the distance between those alleles; the closer the alleles are to each other, the less likely it is that such a reshuffle will occur. Consequently, when selection targets one allele, this automatically results in selection of the other allele as well; through this mechanism, selection can have a strong influence on patterns of variation in the genome.

Selective sweeps occur when an allele becomes more common in a population as a result of positive selection. As the prevalence of one allele increases, linked alleles can also become more common, whether they are neutral or even slightly deleterious. This is called *genetic hitchhiking*. A strong selective sweep results in a region of the genome where the positively selected haplotype (the allele and its neighbours) are in essence the only ones that exist in the population.

Whether a selective sweep has occurred or not can be investigated by measuring linkage disequilibrium, or whether a given haplotype is overrepresented in the population. Normally, genetic recombination results in a reshuffling of the different alleles within a haplotype, and none of the haplotypes will dominate the population. However, during a selective sweep, selection for a specific allele will also result in selection of neighbouring alleles. Therefore, the presence of a block of strong linkage disequilibrium might indicate that there has been a 'recent' selective sweep near the centre of the block, and this can be used to identify sites recently under selection.

Background selection is the opposite of a selective sweep. If a specific site experiences strong and persistent purifying selection, linked variation will tend to be weeded out along with it, producing a region in the genome of low overall variability. Because background selection is a result of deleterious new mutations, which can occur randomly in any haplotype, it does not produce clear blocks of linkage disequilibrium, although with low recombination it can still lead to slightly negative linkage disequilibrium overall.

Competition

In the context of natural selection, competition is an interaction between organisms or species in which the fitness of one is lowered by the presence of another. Limited supply of at least one resource (such as food, water, and territory) used by both can be a factor. Competition; both within and between species is an important topic in ecology, especially community ecology. Competition is one of many interacting biotic and abiotic factors that affect community structure. Competition among members of the same species is known as intraspecific competition, while competition between individuals of different species is known as interspecific competition. Competition is not always straightforward, and can occur in both a direct and indirect fashion.

According to the competitive exclusion principle, species less suited to compete for resources should either adapt or die out, although competitive exclusion is rarely found in natural ecosystems. According to evolutionary theory, this competition within and between species for resources plays a very relevant role in natural selection, however, competition may play less of a role than expansion among larger clades, this is termed the 'Room to Roam' hypothesis.

In evolutionary contexts, competition is related to the concept of r/K selection theory, which relates to the selection of traits which promote success in particular environments. The theory originates from work on island biogeography by the ecologists Robert H. MacArthur and Edward O. Wilson.

In r/K selection theory, selective pressures are hypothesised to drive evolution in one of two stereotyped directions: r- or K-selection. These terms, r and K, are derived from standard ecological algebra, as illustrated in the simple Verhulst equation of population dynamics:

$$\frac{dN}{dt} = rN\left(1 - \frac{N}{K}\right)$$

where r is the growth rate of the population (N), and K is the carrying capacity of its local environmental setting. Typically, r-selected species exploit empty niches, and produce many offspring, each of whom has a relatively low probability of surviving to adulthood. In contrast, K-selected species are strong competitors in crowded niches, and invest more heavily in much fewer offspring, each of whom has a relatively high probability of surviving to adulthood.

Impact of the Idea

Darwin's ideas, along with those of Adam Smith and Karl Marx, had a profound influence on 19th century thought. Perhaps the most radical claim of the theory of evolution through natural selection is that "...elaborately constructed forms, so different from each other, and dependent on each other in so complex a manner, ..." evolved from the simplest forms of life by a few simple principles. This claim inspired some of Darwin's most ardent supporters—and provoked the most profound opposition. The radicalism of natural selection, according to Stephen Jay Gould, lay in its power to "dethrone some of the deepest and most traditional comforts of Western thought." In particular, it challenged long-standing beliefs in such concepts as a special and exalted place for humans in the natural world and a benevolent creator whose intentions were reflected in nature's order and design.

In the words of the philosopher Daniel Dennett, "Darwin's dangerous idea" of evolution by natural selection is a "universal acid," which cannot be kept restricted to any vessel or container, as it soon leaks out, working its way into ever-wider surroundings. Thus, in the last decades, the concept of natural selection has spread from evolutionary biology into virtually all disciplines, including evolutionary computation, quantum Darwinism, evolutionary economics, evolutionary epistemology, evolutionary psychology, and cosmological natural selection. This unlimited applicability has been called universal Darwinism.

Emergence of Natural Selection

How life originated from inorganic matter (abiogensis) remains an unresolved problem in biology. One prominent hypothesis is that life first appeared in the form of short self-replicating RNA polymers. On this view, "life" may have come into existence when RNA chains first experienced the basic conditions, as conceived by Charles Darwin, for natural selection to operate. These conditions are: heritability, variation of type, and competition for limited resources. Fitness of an early RNA replicator (its per capita rate of increase) would likely have been a function of adaptive capacities that were intrinsic (i.e., determined by the nucleotide sequence) and the availability of resources. The three primary adaptive capacities could logically have been: (1) the capacity to replicate with moderate fidelity (giving rise to both heritability and variation of type), (2) the capacity to avoid decay, and (3) the capacity to acquire and process resources. These capacities would have been determined initially by the folded configurations (including those configurations with ribozyme activity) of the RNA replicators that, in turn, would have been encoded in their individual nucleotide sequences. Competitive success among different RNA replicators would have depended on the relative values of their adaptive capacities.

Cell and Molecular Biology

In 1881, Wilhelm Roux, a founder of modern embryology, published *Der Kampf der Theile im Organismus* (*The Struggle of Parts in the Organism*) in which he suggested that the development of an organism results from a Darwinian competition between the parts of the embryo, occurring at all levels, from molecules to organs. In recent years, a modern version of this theory has been proposed by Jean-Jacques Kupiec. According to this cellular Darwinism, Stochasticity at the molecular level generates diversity in cell types whereas cell interactions impose a characteristic order on the developing embryo.

Social and Psychological Theory

The social implications of the theory of evolution by natural selection also became the source of continuing controversy. Friedrich Engels, a German political philosopher and co-originator of the ideology of communism, wrote in 1872 that "Darwin did not know what a bitter satire he wrote on mankind, and especially on his countrymen, when he showed that free competition, the struggle for existence, which the economists celebrate as the highest historical achievement, is the normal state of the *animal kingdom*." Interpretation of natural selection, by Herbert Spencer and Francis Galton as necessarily "progressive," leading to increasing "advances" in intelligence and civilisation, was used as a justification for colonialism and policies of eugenics, as well as to support broader sociopolitical positions now described as social Darwinism. Konrad Lorenz won the Nobel Prize in Physiology or Medicine in 1973 for his analysis of animal behaviour in terms of the role of natural selection (particularly group selection). However, in Germany in 1940, in writings that he subsequently disowned, he used the theory as a justification for policies of the Nazi state. He wrote "... selection for toughness, heroism, and social utility...must be accomplished by some human institution, if mankind, in default of selective factors, is not to be ruined by domestication-induced degeneracy. The racial idea as the basis of our state has already accomplished much in this respect." Others have developed ideas that human societies and culture evolve by mechanisms analogous to those that apply to evolution of species.

More recently, work among anthropologists and psychologists has led to the development of sociobiology and later of evolutionary psychology, a field that attempts to explain features of human psychology in terms of adaptation to the ancestral environment. The most prominent example of evolutionary psychology, notably advanced in the early work of Noam Chomsky and later by Steven Pinker, is the hypothesis that the human brain has adapted to acquire the grammatical rules of natural language. Other aspects of human behaviour and social structures, from specific cultural norms such as incest avoidance to broader patterns such as gender roles, have been hypothesised to have similar origins as adaptations to the early environment in which modern humans evolved. By analogy to the action of natural selection on genes, the concept of memes—"units of cultural transmission," or culture's equivalents of genes undergoing selection and recombination—has arisen, first described in this form by Richard Dawkins in 1976 and subsequently expanded upon by philosophers such as Daniel Dennett as explanations for complex cultural activities, including human consciousness.

Information and Systems Theory

In 1922, Alfred J. Lotka proposed that natural selection might be understood as a physical principle that could be described in terms of the use of energy by a system, a concept that was later developed by Howard Odum as the maximum power principle whereby evolutionary systems with selective advantage maximise the rate of useful energy transformation. Such concepts are sometimes relevant in the study of applied thermodynamics.

The principles of natural selection have inspired a variety of computational techniques, such as "soft" artificial life, that simulate selective processes and can be highly efficient in 'adapting' entities to an environment defined by a specified fitness function. For example, a class of heuristic optimisation algorithms known as genetic algorithms, pioneered by John Henry Holland in the 1970s and expanded upon by David E. Goldberg, identify optimal solutions by simulated reproduction and mutation of a population of solutions defined by an initial probability distribution. Such algorithms are particularly useful when applied to problems whose energy landscape is very rough or has many local minima.

Mode of Natural Selection

Sexual Selection

Sexual selection is a mode of natural selection where members of one biological sex choose mates of the other sex to mate with (intersexual selection), and compete with members of the same sex for access to members of the opposite sex (intrasexual selection). These two forms of selection mean that some individuals have better reproductive success than others within a population, either from being more attractive or preferring more attractive partners to produce offspring. For instance in the breeding season sexual selection in frogs occurs with the males first gathering at the water's edge and making their mating calls: croaking. The females then arrive and choose the males with the deepest croaks and best territories. Generalizing, males benefit from frequent mating and monopolizing access to a group of fertile females. Females have a limited number of offspring they can have and they maximize the return on the energy they invest in reproduction.

Sexual selection creates colourful differences between sexes in Goldie's bird-of-paradise. Male above; female below. Painting by John Gerrard Keulemans (d.1912)

The concept was first articulated by Charles Darwin and Alfred Russel Wallace who described it as driving speciation and that many organisms had evolved features whose function was deleterious to their individual survival, and then developed by Ronald Fisher in the early 20th century. Sexual selection can lead typically males to extreme efforts to demonstrate their fitness to be chosen by females, producing sexual dimorphism in secondary sexual characteristics, such as the ornate plumage of birds such as birds of paradise and peafowl, or the antlers of deer, or the manes of lions, caused by a positive feedback mechanism known as a Fisherian runaway, where the passing on of the desire for a trait in one sex is as important as having the trait in the other sex in producing the runaway effect. Although the sexy son hypothesis indicates that females would prefer male offspring, Fisher's principle explains why the sex ratio is 1:1 almost without exception. Sexual selection is also found in plants and fungi.

The maintenance of sexual reproduction in a highly competitive world is one of the major puzzles in biology given that asexual reproduction can reproduce much more quickly as 50% of offspring are not males, unable to produce offspring themselves. Many non-exclusive hypotheses have been proposed, including the positive impact of an additional form of selection, sexual selection, on the probability of persistence of a species.

In organisms

Sexual selection has been observed to occur in plants, animals and fungi. In certain hermaphroditic snail and slug species of molluscs the throwing of love darts is a form of sexual selection. Certain male insects of the lepidoptera order of insects cement the vaginal pores of their females.

Victorian cartoonists quickly picked up on Darwin's ideas about display in sexual selection.
Here he is fascinated by the apparent steatopygia in the latest fashion.

Illustration from *The Descent of Man* showing the tufted coquette *Lophornis ornatus*: female on left,
ornamented male on right.

- Sexual selection in birds - mammals - humans -scaled reptiles - amphibians - insects - spiders - major histocompatibility complex

Today, biologists say that certain evolutionary traits can be explained by intraspecific competition - competition between members of the same species - distinguishing between competition *before* or *after* sexual intercourse.

- Before copulation, *intrasexual selection* - usually between males - may take the form of *male-to-male combat*. Also, *intersexual selection*, or *mate choice*, occurs when females choose between male mates. Traits selected by male combat are called secondary sexual characteristics (including horns, antlers, etc.), which Darwin described as "weapons", while traits selected by mate (usually female) choice are called "ornaments". Due to their sometimes greatly exaggerated nature, secondary sexual characteristics can prove to be a hindrance to an animal, thereby lowering its chances of survival. For example, the large antlers of a moose are bulky and heavy and slow the creature's flight from predators; they also can become entangled in low-hanging tree branches and shrubs, and undoubtedly have led to the demise of many individuals. Bright colourations and showy ornamentations, such as those seen in many male birds, in addition to capturing the eyes of females, also attract the attention of predators. Some of these traits also represent energetically costly investments for the animals that bear them. Because traits held to be due to sexual selection often conflict with the survival fitness of the individual, the question then arises as to why, in nature, in which survival of the fittest is considered the rule of thumb, such apparent liabilities are allowed to persist. However, one must also consider that intersexual selection can occur with an emphasis on resources that one sex possesses rather than morphological and physiological differences. For example, males of *Euglossa imperialis*, a non-social bee species, form aggregations of territories considered to be leks, to defend fragrant-rich primary territories. The purpose of these aggregations is only facultative, since the more suitable fragrant-rich sites there are, the more habitable territories there are to inhabit, giving females of this species a large selection of males with whom to potentially mate.

- After copulation, male–male competition distinct from conventional aggression may take the form of sperm competition, as described by Parker in 1970. More recently, interest has arisen in *cryptic* female choice, a phenomenon of internally fertilised animals such as mammals and birds, where a female will get rid of a male's sperm without his knowledge.

Finally, sexual conflict is said to occur between breeding partners, sometimes leading to an evolutionary arms race between males and females. Sexual selection can also occur as a product of pheromone release, such as with the Stingless Bee, Trigona corvina.

Female mating preferences are widely recognized as being responsible for the rapid and divergent evolution of male secondary sexual traits. Females of many animal species prefer to mate with males with external ornaments - exaggerated features of morphology such as elaborate sex organs. These preferences may arise when an arbitrary female preference for some aspect of male morphology — initially, perhaps, a result of genetic drift — creates, in due course, selection for males with the appropriate ornament. One interpretation of this is known as the sexy son hypothesis. Alternatively, genes that enable males to develop impressive ornaments or fighting ability may simply show off greater disease resistance or a more efficient metabolism, features that also benefit females. This idea is known as the good genes hypothesis.

In Humans

Darwin conjectured that heritable traits such as beards and hairlessness, significant in the geographical differentiation of human appearance, are results of sexual selection. Geoffrey Miller has hypothesized that many human behaviours not clearly tied to survival benefits, such as humour, music, visual art, verbal creativity, and some forms of altruism, are courtship adaptations that have been favoured through sexual selection. In that view, many human artefacts could be considered subject to sexual selection as part of the extended phenotype, for instance clothing that enhances sexually selected traits. Some argue that the evolution of human intelligence is a sexually selected trait, as it would not confer enough fitness in itself relative to its high maintenance costs.

Darwin and the Development of the Theory

The theory was first proposed by Charles Darwin in *The Origin of Species* (1859) and developed in *The Descent of Man and Selection in Relation to Sex* (1871) because Darwin felt that natural selection alone was unable to account for certain types of non-survival adaptations. He once wrote to a colleague that "The sight of a feather in a peacocks tail, whenever I gaze at it, makes me sick!" His work divided sexual selection into male-male competition and female choice.

... depends, not on a struggle for existence, but on a struggle between the males for possession of the females; the result is not death to the unsuccessful competitor, but few or no offspring.

... when the males and females of any animal have the same general habits ... but differ in structure, colour, or ornament, such differences have been mainly caused by sexual selection.

These views were opposed by Alfred Russel Wallace, though much of his opposition took place after Darwin's death. He argued that male-male competitions were forms of natural selection, but that the "drab" peahen's coloration is itself adaptive as camouflage, and that ascribing mate choice to females was, in his opinion, attributing the ability to judge standards of beauty to animals far too cognitively undeveloped to be capable of aesthetic feeling (such as beetles).

Ronald Fisher

Ronald Fisher

Male long-tailed widowbird

Ronald Fisher, the English statistician and evolutionary biologist developed a number of ideas about sexual selection in his 1930 book *The Genetical Theory of Natural Selection* including the sexy son hypothesis and Fisher's principle. The Fisherian runaway describes how sexual selection accelerates the preference for a specific ornament, causing the preferred trait and female preference for it to increase together in a positive feedback runaway cycle. In a remark that was not widely understood for another 50 years he said:

... plumage development in the male, and sexual preference for such developments in the female, must thus advance together, and so long as the process is unchecked by severe counterselection, will advance with ever-increasing speed. In the total absence of such checks, it is easy to see that the speed of development will be proportional to the development already attained, which will therefore increase with time exponentially, or in geometric progression. —*Ronald Fisher, 1930*

This causes a dramatic increase in both the male's conspicuous feature and in female preference for it, until practical, physical constraints halt further exaggeration. A positive feedback loop is created, producing extravagant physical structures in the non-limiting sex. A classic example of female choice and potential runaway selection is the long-tailed widowbird. While males have long tails that are selected for by female choice, female tastes in tail length are still more extreme with females being attracted to tails longer than those that naturally occur. Fisher understood that female preference for long tails may be passed on genetically, in conjunction with genes for the long tail itself. Long-tailed widowbird offspring of both sexes will inherit both sets of genes, with females expressing their genetic preference for long tails, and males showing off the coveted long tail itself.

Richard Dawkins presents a non-mathematical explanation of the runaway sexual selection process in his book *The Blind Watchmaker*. Females who prefer long tailed males tend to have mothers that chose long-tailed fathers. As a result, they carry both sets of genes in their bodies. That is, genes for long tails and for preferring long tails become linked. The taste for long tails and tail length itself may therefore become correlated, tending to increase together. The more tails lengthen, the more long tails are desired. Any slight initial imbalance between taste and tails may set off an explosion in tail lengths. Fisher corresponded that:

The exponential element, which is the kernel of the thing, arises from the rate of change in hen taste being proportional to the absolute average degree of taste. —*Ronald Fisher, 1932*

The female widow bird will desire to mate with the most attractive long-tailed male so that her progeny, if male, will themselves be attractive to females of the next generation - thereby fathering many offspring who will carry the female's genes. Since the rate of change in preference is proportional to the average taste amongst females, and as females desire to secure the services of the most sexually attractive males, an additive effect is created that, if unchecked, can yield exponential increases in a given taste and in the corresponding desired sexual attribute.

It is important to notice that the conditions of relative stability brought about by these or other means, will be far longer duration than the process in which the ornaments are evolved. In most existing species the runaway process must have been already checked, and we should expect that the more extraordinary developments of sexual plumage are not due like most characters to a long and even course of evolutionary progress, but to sudden spurts of change. —*Ronald Fisher, 1930*

Since Fisher's initial conceptual model of the 'runaway' process, Russell Lande and Peter O'Donald have provided detailed mathematical proofs that define the circumstances under which runaway sexual selection can take place.

Reproductive Success

Extinct Irish elk (*Megaloceros giganteus*). These antlers span 2.7 metres (8.9 ft) and have a mass of 40 kg (88 lb).

The reproductive success of an organism is measured by the number of offspring left behind, and their quality or probable fitness.

The grossest blunder in sexual preference, which we can conceive of an animal making, would be to mate with a species different from its own, and with which hybrids are either infertile, or, through the mixture of instincts and other attributes appropriate to different courses of life, at so serious a disadvantage as to leave no descendants. ... it is no conjecture that a discriminative mechanism exists, variations in which will be capable of giving rise to a similar discrimination within its own species, should such a discrimination become at any time advantageous.

Ronald Fisher, 1930

Individuals in each region most readily attracted to, or excited by, mates of the type there favoured, in contrast to possible mates of the opposite type, will, in fact, be the better represented in future generations, and both the discrimination and the preference will thereby be enhanced. It appears certainly possible that an evolution of sexual preference due to this cause would establish an effective isolation between two differentiated parts of a species, even when geographical and other factors were least favourable to such separation.

Ronald Fisher, 1930

Sexual preference creates a tendency towards assortative mating or homogamy. The general conditions of sexual discrimination appear to be (1) the acceptance of one mate precludes the effective acceptance of alternative mates, and (2) the rejection of an offer will be followed by other offers, either certainly, or at such high chance that the risk of non-occurrence will be smaller than the chance advantage to be gained by selecting a mate.

The conditions determining which sex becomes the more limited resource in intersexual selection can be best understood by way of Bateman's principle which states that "the sex which invests the most in producing offspring becomes a limiting resource over which the other sex will compete", illustrated by the greater nutritional investment of an egg in a zygote, and the limited capacity of females to reproduce; for example in humans a woman can only give birth every ten months whereas in theory a male can become a father every day.

Modern Interpretation

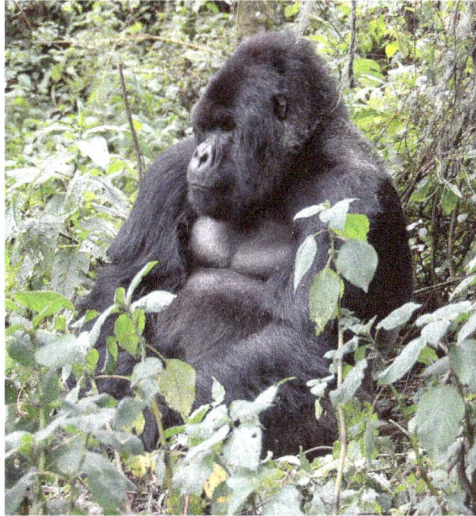

Male mountain gorilla, an example of a tournament species

Flour beetle

Tungara frog

The sciences of evolutionary psychology, human behavioural ecology, and sociobiology study the influence of sexual selection in humans, and are often controversial fields.

Darwin's ideas on sexual selection were met with scepticism by his contemporaries and not considered of great importance in the 20th century, so that in the 1930s biologists decided to include sexual selection as a mode of natural selection. Only in the 21st century have they become more important in biology.

New theories highlight intrinsically useful qualities of such traits. Antlers, horns and the like can be used in physical defence from a predator, and also in competition among males in a tournament species. The winner, which typically becomes the dominant animal in the population, is granted access to females, and therefore increases his reproductive output. Antlers are not the only mechanism that can be used to counteract predation. Predators typically look for the eyes of their prey so they can attack that end of the creature. The conspicuousness of eyespots on many species of butterflies and fishes confuses predators and helps to prevent the prey from suffering serious damage.

Research published in 2015 indicates that sexual selection and the mate choices which "improves population health and protects against extinction, even in the face of genetic stress from high levels of inbreeding" and "ultimately dictates who gets to reproduce their genes into the next generation - so it's a widespread and very powerful evolutionary force." The study involved the flour beetle over a ten-year period where the only changes were in the intensity of sexual selection.

Another, more recently developed, theory, the handicap principle of Amotz Zahavi, Russell Lande and W. D. Hamilton, holds that the fact that the male is able to survive until and through the age of reproduction with such a seemingly maladaptive trait is effectively considered by the female to be a testament to his overall fitness. Such handicaps might prove he is either free of or resistant to disease, or it might demonstrate that this animal possesses more speed or a greater physical strength that is used to combat the troubles brought on by the exaggerated trait.

Zahavi's work spurred a re-examination of the field, which has produced an ever-accelerating number of theories. In 1984, Hamilton and Marlene Zuk introduced the "Bright Male" hypothesis, suggesting that male elaborations might serve as a marker of health, by exaggerating the effects of disease and deficiency. In 1990, Michael Ryan and A.S. Rand, working with the tungara frog, proposed the hypothesis of "Sensory Exploitation", where exaggerated male traits may provide a sensory stimulation that females find hard to resist. Subsequently the theories of the "Gravity Hypothesis" by Jordi Moya-Larano et al. and "Chase Away" by Brett Holland and William R. Rice have also been added. In addition, in the late 1970s Janzen and Mary Willson, noting that male flowers are often larger than female flowers, expanded the field of sexual selection into plants.

In the past few years, the field has exploded to include many additional observations and areas of study, not all of which are clearly included under Darwin's definition of sexual selection. These include cuckoldry, nuptial gifts, sperm competition, infanticide, physical beauty, mating by subterfuge, species isolation mechanisms, male parental care, ambiparental care, mate location, polygamy, and homosexual rape in certain male animals.

Focusing on the effect of sexual conflict, as hypothesized by William Rice, Locke Rowe et Göran Arnvist, Thierry Lodé underlines that the divergence of interest constitutes a key for evolutionary process. Sexual conflict leads to an antagonistic co-evolution in which one sex tends to control the other, resulting in a tug of war. Besides, *the sexual propaganda theory* only argued that mate were opportunistically lead, on the basis of various factors determining the choice such as phenotypic characteristics, apparent vigour of individual, strength of mate signals, trophic resources, territoriality etc. and could explain the maintenance of genetic diversity within populations.

Several workers have brought attention to the fact that elaborated characters that ought to be costly in one way or another for their bearers (e.g., the tails of some species of Xiphophorus fish) do not always appear to have a cost in terms of energetics, performance or even survival. One possible explanation for the apparent lack of costs is that "compensatory traits" have evolved in concert with the sexually selected traits.

As a Toolkit of Natural Selection

Protarchaeopteryx - skull based on *Incisivosaurus* and wings on *Caudipteryx*

Sexual selection may explain how certain characteristics (such as feathers) had distinct survival value at an early stage in their evolution.

Geoffrey Miller proposes that sexual selection might have contributed by creating evolutionary modules such as *Archaeopteryx* feathers as sexual ornaments, at first. The earliest proto-birds such as China's *Protarchaeopteryx*, discovered in the early 1990s, had well-developed feathers but no sign of the top/bottom asymmetry that gives wings lift. Some have suggested that the feathers served as insulation, helping females incubate their eggs. But perhaps the feathers served as the kinds of sexual ornaments still common in most bird species, and especially in birds such as peacocks and birds-of-paradise today. If proto-bird courtship displays combined displays of forelimb feathers with energetic jumps, then the transition from display to aerodynamic functions could have been relatively smooth.

Sexual selection sometimes generates features that may help cause a species' extinction, as has been suggested for the giant antlers of the Irish elk (*Megaloceros giganteus*) that became extinct in Pleistocene Europe. However, sexual selection can also do the opposite, driving species divergence - sometimes through elaborate changes in genitalia - such that new species emerge.

Sexual Dimorphism

Sex differences directly related to reproduction and serving no direct purpose in courtship are called primary sexual characteristics. Traits amenable to sexual selection, which give an organism an advantage over its rivals (such as in courtship) without being directly involved in reproduction, are called secondary sex characteristics.

In most sexual species the males and females have different equilibrium strategies, due to a difference in relative investment in producing offspring. As formulated in Bateman's principle, females have a greater initial investment in producing offspring (pregnancy in mammals or the production of the egg in birds and reptiles), and this difference in initial investment creates differences in variance in expected reproductive success and bootstraps the sexual selection processes. Classic examples of reversed sex-role species include the pipefish, and Wilson's phalarope. Also, unlike a female, a male (except in monogamous species) has some uncertainty about whether or not he is the true parent of a child, and so will be less interested in spending his energy helping to raise offspring that may or may not be related to him. As a result of these factors, males are typically more willing to mate than females, and so females are typically the ones doing the choosing (except in cases of forced copulations, which can occur in certain species of primates, ducks, and others). The effects of sexual selection are thus held to typically be more pronounced in males than in females.

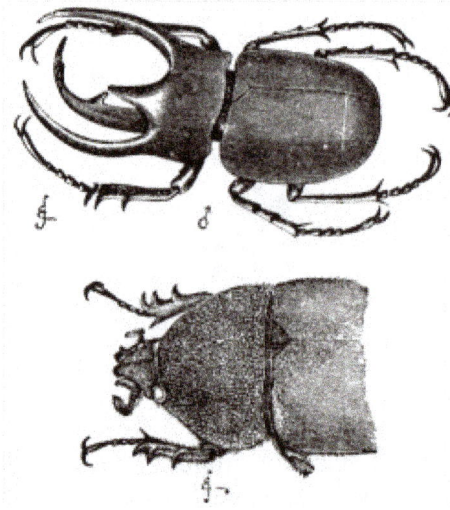

The rhinoceros beetle is a classic case of sexual dimorphism. Plate from Darwin's *Descent of Man*, male at top, female at bottom

Differences in secondary sexual characteristics between males and females of a species are referred to as sexual dimorphisms. These can be as subtle as a size difference (sexual size dimorphism, often abbreviated as SSD) or as extreme as horns and colour patterns. Sexual dimorphisms abound in nature. Examples include the possession of antlers by only male deer, the brighter coloration of many male birds in comparison with females of the same species, or even more distinct differences in basic morphology, such as the drastically increased eye-span of the male stalk-eyed fly. The peacock, with its elaborate and colourful tail feathers, which the peahen lacks, is often referred to as perhaps the most extraordinary example of a dimorphism. Male and female black-throated blue warblers and Guianan cock-of-the-rocks also differ radically in their plumage. Early naturalists even believed the females to be a separate species. The largest sexual size dimorphism in vertebrates is the shell dwelling cichlid fish *Neolamprologus callipterus* in which males are up to 30 times the size of females. Many other fish such as guppies also exhibit sexual dimorphism. Extreme sexual size dimorphism, with females larger than males, is quite common in spiders and birds of prey.

Heredity

Heredity is the genetic information passing for traits from parents to their offspring, either through asexual reproduction or sexual reproduction. This is the process by which an offspring cell or organism acquires or becomes predisposed to the characteristics of its parent cell or organism. Through heredity, variations exhibited by individuals can accumulate and cause some species to evolve through the natural selection of specific phenotype traits. The study of heredity in biology is called genetics, which includes the field of epigenetics.

Overview

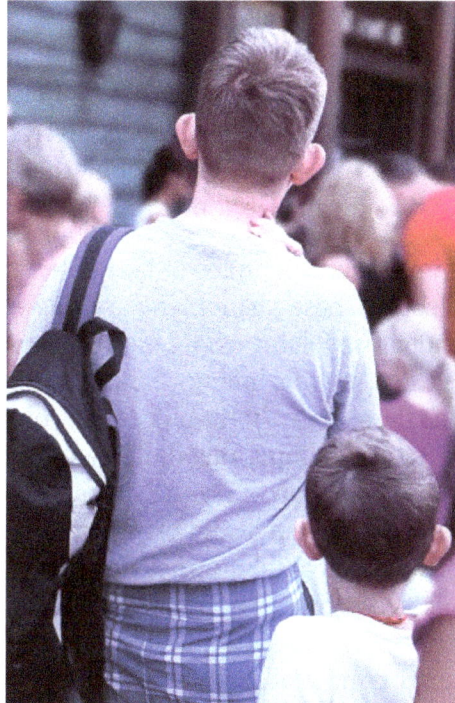

Heredity of phenotypic traits: Father and son with prominent ears and crowns.

In humans, eye color is an example of an inherited characteristic: an individual might inherit the "brown-eye trait" from one of the parents. Inherited traits are controlled by genes and the complete set of genes within an organism's genome is called its genotype.

The complete set of observable traits of the structure and behavior of an organism is called its phenotype. These traits arise from the interaction of its genotype with the environment. As a result, many aspects of an organism's phenotype are not inherited. For example, suntanned skin comes from the interaction between a person's phenotype and sunlight; thus, suntans are not passed on to people's children. However, some people tan more easily than others, due to differences in their genotype: a striking example is people with the inherited trait of albinism, who do not tan at all and are very sensitive to sunburn.

Heritable traits are known to be passed from one generation to the next via DNA, a molecule that encodes genetic information. DNA is a long polymer that incorporates four types of bases, which are interchangeable. The sequence of bases along a particular DNA molecule specifies the genetic information: this is comparable to a sequence of letters spelling out a passage of text. Before a cell divides through mitosis, the DNA is copied, so that each of the resulting two cells will inherit the DNA sequence. A portion of a DNA molecule that specifies a single functional unit is called a gene; different genes have different sequences of bases. Within cells, the long strands of DNA form condensed structures called chromosomes. Organisms inherit genetic material from their parents in the form of homologous chromosomes, containing a unique combination of DNA sequences that code for genes. The specific location of a DNA sequence within a chromosome is known as a locus. If the DNA sequence at a particular locus varies between individuals, the different forms of this sequence are called alleles. DNA sequences can change through mutations, producing new alleles. If a mutation occurs within a gene, the new allele may affect the trait that the gene controls, altering the phenotype of the organism.

However, while this simple correspondence between an allele and a trait works in some cases, most traits are more complex and are controlled by multiple interacting genes within and among organisms. Developmental biologists suggest that complex interactions in genetic networks and communication among cells can lead to heritable variations that may underlie some of the mechanics in developmental plasticity and canalization.

Recent findings have confirmed important examples of heritable changes that cannot be explained by direct agency of the DNA molecule. These phenomena are classed as epigenetic inheritance systems that are causally or independently evolving over genes. Research into modes and mechanisms of epigenetic inheritance is still in its scientific infancy, however, this area of research has attracted much recent activity as it broadens the scope of heritability and evolutionary biology in general. DNA methylation marking chromatin, self-sustaining metabolic loops, gene silencing by RNA interference, and the three dimensional conformation of proteins (such as prions) are areas where epigenetic inheritance systems have been discovered at the organismic level. Heritability may also occur at even larger scales. For example, ecological inheritance through the process of niche construction is defined by the regular and repeated activities of organisms in their environment. This generates a legacy of effect that modifies and feeds back into the selection regime of subsequent generations. Descendants inherit genes plus environmental characteristics generated by the ecological actions of ancestors. Other examples of heritability in evolution that are not under the direct control of genes include the inheritance of cultural traits, group heritability, and symbiogenesis. These examples of heritability that operate above the gene are covered broadly under the title of multilevel or hierarchical selection, which has been a subject of intense debate in the history of evolutionary science.

Relation to Theory of Evolution

When Charles Darwin proposed his theory of evolution in 1859, one of its major problems was the lack of an underlying mechanism for heredity. Darwin believed in a mix of blending inheritance and the inheritance of acquired traits (pangenesis). Blending inheritance would lead to uniformity across populations in only a few generations and then would remove variation from a population on which natural selection could act. This led to Darwin adopting some Lamarckian ideas in later editions of *On the Origin of Species* and his later biological works. Darwin's primary approach to heredity was to outline how it appeared to work (noticing that traits that were not expressed explicitly in the parent at the time of reproduction could be inherited, that certain traits could be sex-linked, etc.) rather than suggesting mechanisms.

Darwin's initial model of heredity was adopted by, and then heavily modified by, his cousin Francis Galton, who laid the framework for the biometric school of heredity. Galton found no evidence to support the aspects of Darwin's pangenesis model, which relied on acquired traits.

The inheritance of acquired traits was shown to have little basis in the 1880s when August Weismann cut the tails off many generations of mice and found that their offspring continued to develop tails.

History

Scientists in Antiquity had a variety of ideas about heredity: Theophrastus proposed that male flowers caused female flowers to ripen; Hippocrates speculated that "seeds" were produced by various body parts and transmitted to offspring at the time of conception; and Aristotle thought that male and female semen mixed at conception. Aeschylus, in 458 BC, proposed the male as the parent, with the female as a "nurse for the young life sown within her".

Ancient understandings of heredity transitioned to two debated doctrines in the 18th century. The Doctrine of Epigenesis and the Doctrine of Preformation were two distinct views of the understanding of heredity. The Doctrine of Epigenesis, originated by Aristotle, claimed that an embryo continually develops. The modifications of the parent's traits are passed off to an embryo during its lifetime. The foundation of this doctrine was based on the theory of inheritance of acquired traits. In direct opposition, the Doctrine of Preformation claimed that "like generates like" where the germ would evolve to yield offspring similar to the parents. The Preformationist view believed procreation was an act of revealing what had been created long before. However, this was disputed by the creation of the cell theory in the 19th century, where the fundamental unit of life is the cell, and not some preformed parts of an organism. Various hereditary mechanisms, including blending inheritance were also envisaged without being properly tested or quantified, and were later disputed. Nevertheless, people were able to develop domestic breeds of animals as well as crops through artificial selection. The inheritance of acquired traits also formed a part of early Lamarckian ideas on evolution.

During the 18th century, Dutch microscopist Antonie van Leeuwenhoek (1632–1723) discovered "animalcules" in the sperm of humans and other animals. Some scientists speculated they saw a "little man" (homunculus) inside each sperm. These scientists formed a school of thought known as the "spermists". They contended the only contributions of the female to the next generation were the womb in which the homunculus grew, and prenatal influences of the womb. An opposing school of thought, the ovists, believed that the future human was in the egg, and that sperm merely stimulated the growth of the egg. Ovists thought women carried eggs containing boy and girl children, and that the gender of the offspring was determined well before conception.

Gregor Mendel: Father of Genetics

The idea of particulate inheritance of genes can be attributed to the Moravian monk Gregor Mendel who published his work on pea plants in 1865. However, his work was not widely known and was rediscovered in 1901. It was initially assumed the Mendelian inheritance only accounted for large (qualitative) differences, such as those seen by Mendel in his pea plants—and the idea of additive effect of (quantitative) genes was not realised until R. A. Fisher's (1918) paper, "The Correlation Between Relatives on the Supposition of Mendelian Inheritance" Mendel's overall contribution gave scientists a useful overview that traits were inheritable. His pea plant demonstration became the foundation of the study of Mendelian Traits. These traits can be traced on a single locus.

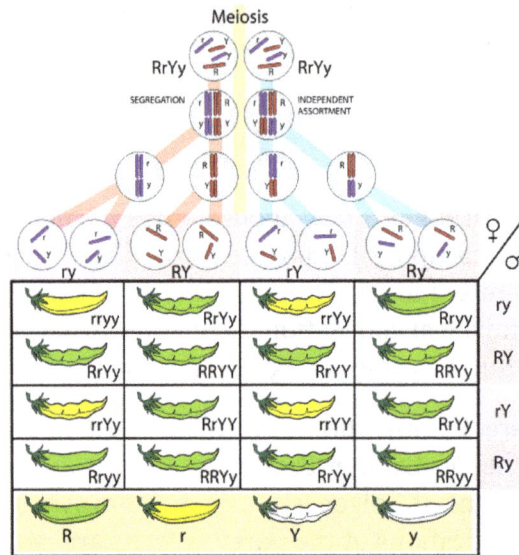

Table showing how the genes exchange according to segregation or independent assortment during meiosis and how this translates into Mendel's laws

Modern Development of Genetics and Heredity

In the 1930s, work by Fisher and others resulted in a combination of Mendelian and biometric schools into the modern evolutionary synthesis. The modern synthesis bridged the gap between experimental geneticists and naturalists; and between both and palaeontologists, stating that:

1. All evolutionary phenomena can be explained in a way consistent with known genetic mechanisms and the observational evidence of naturalists.

2. Evolution is gradual: small genetic changes, recombination ordered by natural selection. Discontinuities amongst species (or other taxa) are explained as originating gradually through geographical separation and extinction (not saltation).

3. Selection is overwhelmingly the main mechanism of change; even slight advantages are important when continued. The object of selection is the phenotype in its surrounding environment. The role of genetic drift is equivocal; though strongly supported initially by Dobzhansky, it was downgraded later as results from ecological genetics were obtained.

4. The primacy of population thinking: the genetic diversity carried in natural populations is a key factor in evolution. The strength of natural selection in the wild was greater than expected; the effect of ecological factors such as niche occupation and the significance of barriers to gene flow are all important.

The idea that speciation occurs after populations are reproductively isolated has been much debated. In plants, polyploidy must be included in any view of speciation. Formulations such as 'evolution consists primarily of changes in the frequencies of alleles between one generation and another' were proposed rather later. The traditional view is that developmental biology ('evo-devo') played little part in the synthesis, but an account of Gavin de Beer's work by Stephen Jay Gould suggests he may be an exception.

Almost all aspects of the synthesis have been challenged at times, with varying degrees of success. There is no doubt, however, that the synthesis was a great landmark in evolutionary biology. It cleared up many confusions, and was directly responsible for stimulating a great deal of research in the post-World War II era.

Trofim Lysenko however caused a backlash of what is now called Lysenkoism in the Soviet Union when he emphasised Lamarckian ideas on the inheritance of acquired traits. This movement affected agricultural research and led to food shortages in the 1960s and seriously affected the USSR.

Common Genetic Disorders

- Down syndrome
- sickle cell disease
- Phenylketonuria (PKU)
- Haemophilia

Types

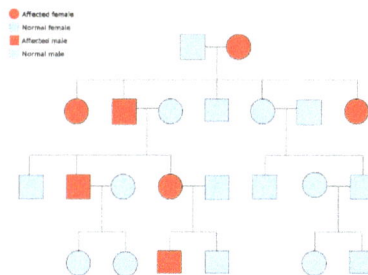

An example pedigree chart of an autosomal dominant disorder.

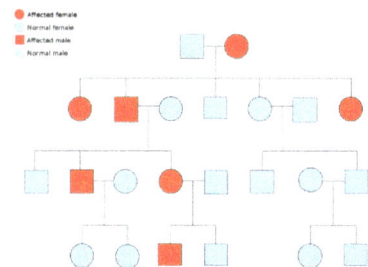

An example pedigree chart of an autosomal recessive disorder.

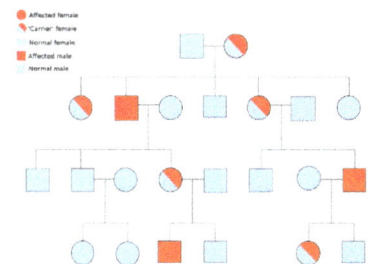

An example pedigree chart of a sex-linked disorder (the gene is on the X chromosome)

Dominant and Recessive Alleles

An allele is said to be dominant if it is always expressed in the appearance of an organism (phenotype) provided that at least one copy of it is present. For example, in peas the allele for green pods, *G*, is dominant to that for yellow pods, *g*. Thus pea plants with the pair of alleles **either** *GG* (homozygote) **or** *Gg* (heterozygote) will have green pods. The allele for yellow pods is recessive. The effects of this allele are only seen when it is present in both chromosomes, *gg* (homozygote).

The description of a mode of biological inheritance consists of three main categories:

1. Number of involved loci

- Monogenetic (also called "simple") – one locus

- Oligogenetic – few loci

- Polygenetic – many loci

2. Involved chromosomes

- Autosomal – loci are not situated on a sex chromosome

- Gonosomal – loci are situated on a sex chromosome

  o X-chromosomal – loci are situated on the X-chromosome (the more common case)

  o Y-chromosomal – loci are situated on the Y-chromosome

- Mitochondrial – loci are situated on the mitochondrial DNA

3. Correlation genotype–phenotype

- Dominant

- Intermediate (also called "codominant")

- Recessive

- Overdominant

- Underdominant

These three categories are part of every exact description of a mode of inheritance in the above order. In addition, more specifications may be added as follows:

4. Coincidental and environmental interactions

- Penetrance

  o Complete

  o Incomplete (percentual number)

- Expressivity

  o Invariable

       ○   Variable

- Heritability (in polygenetic and sometimes also in oligogenetic modes of inheritance)

- Maternal or paternal imprinting phenomena

5. Sex-linked interactions

- Sex-linked inheritance (gonosomal loci)

- Sex-limited phenotype expression (e.g., cryptorchism)

- Inheritance through the maternal line (in case of mitochondrial DNA loci)

- Inheritance through the paternal line (in case of Y-chromosomal loci)

6. Locus–locus interactions

- Epistasis with other loci (e.g., overdominance)

- Gene coupling with other loci

- Homozygotous lethal factors

- Semi-lethal factors

Determination and description of a mode of inheritance is achieved primarily through statistical analysis of pedigree data. In case the involved loci are known, methods of molecular genetics can also be employed.

Genotype

Here the relation between genotype and phenotype is illustrated, using a Punnett square, for the character of petal colour in a pea plant. The letters B and b represent genes for colour and the pictures show the resultant flowers.

The genotype is the part (DNA sequence) of the genetic makeup of a cell, and therefore of an organism or individual, which determines a specific characteristic (phenotype) of that cell/organism/individual. Genotype is one of three factors that determine phenotype, the other two being inherited epigenetic factors, and non-inherited environmental factors. DNA mutations which are acquired rather than inherited, such as cancer mutations, are not part of the individual's genotype; hence, scientists and physicians sometimes talk for example about the (geno)type of a particular cancer, that is the genotype of the disease as distinct from the diseased.

An example of how genotype determines a characteristic is petal color in a pea plant.

Genotypic and Genomic Sequence

The genotype of an organism is the inherited map it carries within its genetic code. Not all organisms with the same genotype look or act the same way because appearance and behavior are modified by environmental and developmental conditions. Likewise, not all organisms that look alike necessarily have the same genotype. One's genotype differs subtly from one's genomic sequence. A sequence is an absolute measure of base composition of an individual, or a representative of a species or group; a genotype typically implies a measurement of how an individual *differs* or is specialized within a group of individuals or a species. So typically, one refers to an individual's genotype with regard to a particular gene of interest and, in polyploid individuals, it refers to what combination of alleles the individual carries. The genetic constitution of an organism is referred to as its genotype, such as the letters Bb. (B - dominant genotype and b - recessive genotype).

Genotype and Phenotype

Any given gene will usually cause an observable change in an organism, known as the phenotype. The terms genotype and phenotype are distinct for at least two reasons:

1. To distinguish the source of an observer's knowledge (one can know about genotype by observing DNA; one can know about phenotype by observing outward appearance of an organism).

2. Genotype and phenotype are not always directly correlated. Some genes only express a given phenotype in certain environmental conditions. Conversely, some phenotypes could be the result of multiple genotypes. The genotype is commonly mixed up with the phenotype which describes the end result of both the genetic and the environmental factors giving the observed expression (e.g. blue eyes, hair color, or various hereditary diseases).

A simple example to illustrate genotype as distinct from phenotype is the flower colour in pea plants. There are three available genotypes, PP (homozygous dominant), Pp (heterozygous), and pp (homozygous recessive). All three have different genotypes but the first two have the same phenotype (purple) as distinct from the third (white).

A more technical example to illustrate genotype is the single nucleotide polymorphism or SNP. A SNP occurs when corresponding sequences of DNA from different individuals differ at one DNA base, for example where the sequence AAGCCTA changes to AAGCTTA. This contains two alleles : C and T. SNPs typically have three genotypes, denoted generically AA Aa and aa. In the example above, the three genotypes would be CC, CT and TT. Other types of genetic marker, such as microsatellites, can have more than two alleles, and thus many different genotypes.

Genotype and Mendelian Inheritance

The distinction between genotype and phenotype is commonly experienced when studying family patterns for certain hereditary diseases or conditions, for example, haemophilia. Due to the diploidy of humans (and most animals), there are two alleles for any given gene. These alleles can be the same (homozygous) or different (heterozygous), depending on the individual. With a dominant allele, the offspring is guaranteed to inherit the trait in question irrespective of the second allele.

In the case of an albino with a recessive allele (aa), the phenotype depends upon the other allele (Aa, aA, aa or AA). An affected person mating with a heterozygous individual (Aa or aA, also carrier) there is a 50-50 chance the offspring will be albino's phenotype. If a heterozygote mates with another heterozygote, there is 75% chance passing the gene on and only a 25% chance that the gene will be displayed. A homozygous dominant (AA) individual has a normal phenotype and no risk of abnormal offspring. A homozygous recessive individual has an abnormal phenotype and is guaranteed to pass the abnormal gene onto offspring.

In the case of haemophilia, it is sex-linked thus only carried on the X chromosome. Only females can be a carrier in which the abnormality is not displayed. This woman has a normal phenotype, but runs a 50-50 chance, with an unaffected partner, of passing her abnormal gene on to her offspring. If she mated with a man with haemophilia (another carrier) there would be a 75% chance of passing on the gene.

Genotype and Mathematics

Inspired by the biological concept and usefulness of genotypes, computer science employs simulated phenotypes in genetic programming and evolutionary algorithms. Such techniques can help evolve mathematical solutions to certain types of otherwise difficult problems.

Determining Genotype

Genotyping is the process of elucidating the genotype of an individual with a biological assay. Also known as a *genotypic assay*, techniques include PCR, DNA fragment analysis, allele specific oligonucleotide (ASO) probes, DNA sequencing, and nucleic acid hybridization to DNA microarrays or beads. Several common genotyping techniques include restriction fragment length polymorphism (*RFLP*), terminal restriction fragment length polymorphism (*t-RFLP*), amplified fragment length polymorphism (*AFLP*), and multiplex ligation-dependent probe amplification (*MLPA*).

DNA fragment analysis can also be used to determine such disease causing genetics aberrations as microsatellite instability (*MSI*), *trisomy* or aneuploidy, and loss of heterozygosity (*LOH*). MSI and LOH in particular have been associated with cancer cell genotypes for colon, breast and cervical cancer.

The most common chromosomal aneuploidy is a trisomy of chromosome 21 which manifests itself as Down syndrome. Current technological limitations typically allow only a fraction of an individual's genotype to be determined efficiently.

References

- Fisher, RA (1930) The Genetical Theory of Natural Selection. Oxford University Press, ISBN 0-19-850440-3, Chapter 6 Memeoid.net

- Miller, GF (2000) The Mating Mind: How sexual choice shaped the evolution of human nature. Heinemann, London. ISBN 0-434-00741-2

- Griffiths, Anthony, J. F.; Wessler, Susan R.; Carroll, Sean B.; Doebley J (2012). Introduction to Genetic Analysis (10th ed.). New York: W. H. Freeman and Company. p. 3. ISBN 978-1-4292-2943-2.

- Jablonka, E.; Lamb, M. (2005). Evolution in four dimensions: Genetic, epigenetic, behavioural, and symbolic. MIT Press. ISBN 0-262-10107-6.

- Griffiths, Anthony, J. F.; Wessler, Susan R.; Carroll, Sean B.; Doebley, John (2012). Introduction to Genetic Analysis (10th ed.). New York: W. H. Freeman and Company. p. 14. ISBN 978-1-4292-2943-2.

- Lipton, Bruce H. (2008). The Biology of Belief: Unleashing the Power of Consciousness, Matter and Miracles. Hay House, Inc. p. 12. ISBN 9781401923440.

- Hipócrates (1981). Hippocratic Treatises: On Generation - Nature of the Child - Diseases Ic. Walter de Gruyter. p. 6. ISBN 9783110079036.

- Gottlieb, Gilbert (2001). Individual Development and Evolution: The Genesis of Novel Behavior. Psychology Press. p. 4. ISBN 9781410604422.

- Carlson, Neil and et al. Psychology the Science of Behavior, p. 206. Pearson Canada, United States of America. ISBN 978-0-205-64524-4.

- Wright, Sewall (1932). "The roles of mutation, inbreeding, crossbreeding and selection in evolution". Proceedings of the VI International Congress of Genetrics. 1: 356–366. Retrieved 2015-08-02.

- Loewe, Laurence (2008). "Negative Selection". Nature Education. Cambridge, MA: Nature Publishing Group. OCLC 310450541. Retrieved 2015-08-03.

- Gould, Stephen Jay (June 12, 1997). "Darwinian Fundamentalism". The New York Review of Books. New York: Rea S. Hederman. 44 (10). ISSN 0028-7504. Retrieved 2015-08-03.

- Kupiec, Jean-Jacques (May 3, 2010). "Cellular Darwinism (stochastic gene expression in cell differentiation and embryo development)". SciTopics. Archived from the original on 2010-08-04. Retrieved 2015-08-11.

- Lawrence, Cera R. (2008). Hartsoeker's Homunculus Sketch from Essai de Dioptrique. Embryo Project Encyclopedia. ISSN 1940-5030. Retrieved March 26, 2013.

- Shoag J; et al. (Jan 2013). "PGC-1 coactivators regulate MITF and the tanning response". Mol Cell. 49 (1): 145–57. doi:10.1016/j.molcel.2012.10.027. PMID 23201126.

- "Aristotle's Biology - 5.2. From Inquiry to Understanding; from hoti to dioti.". Stanford University. Feb 15, 2006. Retrieved March 26,

Evolution: An Overview

The change in the characteristics of a biological population over a period of years is called evolution whereas the process of change in the composition of sequences such as DNA, RNS and proteins is known as molecular evolution. This chapter on evolution offers an insightful focus, keeping in mind the complex subject matter.

Evolution

Evolution is change in the heritable traits of biological populations over successive generations. Evolutionary processes give rise to biodiversity at every level of biological organisation, including the levels of species, individual organisms, and molecules.

All life on Earth shares a common ancestor known as the last universal common ancestor (LUCA), which lived approximately 3.5–3.8 billion years ago, although a study in 2015 found "remains of biotic life" from 4.1 billion years ago in ancient rocks in Western Australia. In July 2016, scientists reported identifying a set of 355 genes from the LUCA of all organisms living on Earth.

Repeated formation of new species (speciation), change within species (anagenesis), and loss of species (extinction) throughout the evolutionary history of life on Earth are demonstrated by shared sets of morphological and biochemical traits, including shared DNA sequences. These shared traits are more similar among species that share a more recent common ancestor, and can be used to reconstruct a biological "tree of life" based on evolutionary relationships (phylogenetics), using both existing species and fossils. The fossil record includes a progression from early biogenic graphite, to microbial mat fossils, to fossilized multicellular organisms. Existing patterns of biodiversity have been shaped both by speciation and by extinction. More than 99 percent of all species that ever lived on Earth are estimated to be extinct. Estimates of Earth's current species range from 10 to 14 million, of which about 1.2 million have been documented. More recently, in May 2016, scientists reported that 1 trillion species are estimated to be on Earth currently with only one-thousandth of one percent described.

In the mid-19th century, Charles Darwin formulated the scientific theory of evolution by natural selection, published in his book *On the Origin of Species* (1859). Evolution by natural selection is a process demonstrated by the observation that more offspring are produced than can possibly survive, along with three facts about populations: 1) traits vary among individuals with respect to morphology, physiology, and behaviour (phenotypic variation), 2) different traits confer different rates of survival and reproduction (differential fitness), and 3) traits can be passed from generation to generation (heritability of fitness). Thus, in successive generations members of a population are replaced by progeny of parents better adapted to survive and reproduce in the biophysical environment in which natural selection takes place. This teleonomy is the quality whereby the process of natural selection creates and preserves traits that are seemingly fitted for the functional roles they perform. Natural selection, including sexual selection, is the only known cause of adaptation but not the only known cause of evolution. Other, nonadaptive evolutionary processes include mutation, genetic drift and gene migration.

In the early 20th century the modern evolutionary synthesis integrated classical genetics with Darwin's theory of evolution by natural selection through the discipline of population genetics. The importance of natural selection as a cause of evolution was accepted into other branches of biology. Moreover, previously held notions about evolution, such as orthogenesis, evolutionism, and other beliefs about innate "progress" within the largest-scale trends in evolution, became obsolete scientific theories. Scientists continue to study various aspects of evolutionary biology by forming and testing hypotheses, constructing mathematical models of theoretical biology and biological theories, using observational data, and performing experiments in both the field and the laboratory.

In terms of practical application, an understanding of evolution has been instrumental to developments in numerous scientific and industrial fields, including agriculture, human and veterinary medicine, and the life sciences in general. Discoveries in evolutionary biology have made a significant impact not just in the traditional branches of biology but also in other academic disciplines, including biological anthropology, and evolutionary psychology. Evolutionary computation, a subfield of artificial intelligence, involves the application of Darwinian principles to problems in computer science.

History of Evolutionary Thought

Biologist and statistician Ronald Fisher

In 1842, Charles Darwin penned his first sketch of what became *On the Origin of Species*.

The proposal that one type of organism could descend from another type goes back to some of the first pre-Socratic Greek philosophers, such as Anaximander and Empedocles. Such proposals survived into Roman times. The poet and philosopher Lucretius followed Empedocles in his masterwork *De rerum natura* (*On the Nature of Things*). In contrast to these materialistic views, Aristotle considered all natural things, not only living things, as being imperfect actualisations of different fixed natural possibilities, known as "forms," "ideas," or (in Latin translations) "*species*." This was part of his teleological understanding of nature in which all things have an intended role to play in a divine cosmic order. Variations of this idea became the standard understanding of the Middle Ages and were integrated into Christian learning, but Aristotle did not demand that real types of organisms always correspond one-for-one with exact metaphysical forms and specifically gave examples of how new types of living things could come to be.

In the 17th century, the new method of modern science rejected Aristotle's approach. It sought explanations of natural phenomena in terms of physical laws that were the same for all visible things and that did not require the existence of any fixed natural categories or divine cosmic order. However, this new approach was slow to take root in the biological sciences, the last bastion of the concept of fixed natural types. John Ray applied one of the previously more general terms for fixed natural types, "species," to plant and animal types, but he strictly identified each type of living thing as a species and proposed that each species could be defined by the features that perpetuated themselves generation after generation. The biological classification introduced by Carl Linnaeus in 1735 explicitly recognized the hierarchical nature of species relationships, but still viewed species as fixed according to a divine plan.

Other naturalists of this time speculated on the evolutionary change of species over time according to natural laws. In 1751, Pierre Louis Maupertuis wrote of natural modifications occurring during reproduction and accumulating over many generations to produce new species. Georges-Louis Leclerc, Comte de Buffon suggested that species could degenerate into different organisms, and Erasmus Darwin proposed that all warm-blooded animals could have descended from a single microorganism (or "filament"). The first full-fledged evolutionary scheme was Jean-Baptiste Lamarck's "transmutation" theory of 1809, which envisaged spontaneous generation continually producing simple forms of life that developed greater complexity in parallel lineages with an inherent progressive tendency, and postulated that on a local level these lineages adapted to the environment by inheriting changes caused by their use or disuse in parents. (The latter process was later called Lamarckism.) These ideas were condemned by established naturalists as speculation lacking empirical support. In particular, Georges Cuvier insisted that species were unrelated and fixed, their similarities reflecting divine design for functional needs. In the meantime, Ray's ideas of benevolent design had been developed by William Paley into the *Natural Theology or Evidences of the Existence and Attributes of the Deity* (1802), which proposed complex adaptations as evidence of divine design and which was admired by Charles Darwin.

The crucial break from the concept of constant typological classes or types in biology came with the theory of evolution through natural selection, which was formulated by Charles Darwin in terms of variable populations. Partly influenced by *An Essay on the Principle of Population* (1798) by Thomas Robert Malthus, Darwin noted that population growth would lead to a "struggle for existence" in which favorable variations prevailed as others perished. In each generation, many offspring fail to survive to an age of reproduction because of limited resources. This could explain the diversity of plants and animals from a common ancestry through the working of natural laws in the same way for all types of organism. Darwin developed his theory of "natural selection" from 1838 onwards and was writing up his "big book" on the subject when Alfred Russel Wallace sent him a version of virtually the same theory in 1858. Their separate papers were presented together at a 1858 meeting of the Linnean Society of London. At the end of 1859, Darwin's publication of his "abstract" as *On the Origin of Species* explained natural selection in detail and in a way that led to an increasingly wide acceptance of concepts of evolution. Thomas Henry Huxley applied Darwin's ideas to humans, using paleontology and comparative anatomy to provide strong evidence that humans and apes shared a common ancestry. Some were disturbed by this since it implied that humans did not have a special place in the universe.

Precise mechanisms of reproductive heritability and the origin of new traits remained a mystery. Towards this end, Darwin developed his provisional theory of pangenesis. In 1865, Gregor Mendel reported that traits were inherited in a predictable manner through the independent assortment and segregation of elements (later known as genes). Mendel's laws of inheritance eventually supplanted most of Darwin's pangenesis theory. August Weismann made the important distinction between germ cells that give rise to gametes (such as sperm and egg cells) and the somatic cells of the body, demonstrating that heredity passes through the germ line only. Hugo de Vries connected Darwin's pangenesis theory to Weismann's germ/soma cell distinction and proposed that Darwin's pangenes were concentrated in the cell nucleus and when expressed they could move into the cytoplasm to change the cells structure. De Vries was also one of the researchers who made Mendel's work well-known, believing that Mendelian traits corresponded to the transfer of heritable variations along the germline. To explain how new variants originate, de Vries developed a mutation theory that led to a temporary rift between those who accepted Darwinian evolution and biometricians who allied with de Vries. In the 1930s, pioneers in the field of population genetics, such as Ronald Fisher, Sewall Wright and J. B. S. Haldane set the foundations of evolution onto a robust statistical philosophy. The false contradiction between Darwin's theory, genetic mutations, and Mendelian inheritance was thus reconciled.

In the 1920s and 1930s a modern evolutionary synthesis connected natural selection, mutation theory, and Mendelian inheritance into a unified theory that applied generally to any branch of biology. The modern synthesis was able to explain patterns observed across species in populations, through fossil transitions in palaeontology, and even complex cellular mechanisms in developmental biology. The publication of the structure of DNA by James Watson and Francis Crick in 1953 demonstrated a physical mechanism for inheritance. Molecular biology improved our understanding of the relationship between genotype and phenotype. Advancements were also made in phylogenetic systematics, mapping the transition of traits into a comparative and testable framework through the publication and use of evolutionary trees. In 1973, evolutionary biologist Theodosius Dobzhansky penned that "nothing in biology makes sense except in the light of evolution," because it has brought to light the relations of what first seemed disjointed facts in natural history into a coherent explanatory body of knowledge that describes and predicts many observable facts about life on this planet.

Since then, the modern synthesis has been further extended to explain biological phenomena across the full and integrative scale of the biological hierarchy, from genes to species. This extension, known as evolutionary developmental biology and informally called "evo-devo," emphasises how changes between generations (evolution) acts on patterns of change within individual organisms (development).

Heredity

Evolution in organisms occurs through changes in heritable traits—the inherited characteristics of an organism. In humans, for example, eye colour is an inherited characteristic and an individual might inherit the "brown-eye trait" from one of their parents. Inherited traits are controlled by genes and the complete set of genes within an organism's genome (genetic material) is called its genotype.

The complete set of observable traits that make up the structure and behaviour of an organism is called its phenotype. These traits come from the interaction of its genotype with the environment. As a result, many aspects of an organism's phenotype are not inherited. For example, suntanned skin comes from the interaction between a person's genotype and sunlight; thus, suntans are not passed on to people's children. However, some people tan more easily than others, due to differences in genotypic variation; a striking example are people with the inherited trait of albinism, who do not tan at all and are very sensitive to sunburn.

Heritable traits are passed from one generation to the next via DNA, a molecule that encodes genetic information. DNA is a long biopolymer composed of four types of bases. The sequence of bases along a particular DNA molecule specify the genetic information, in a manner similar to a sequence of letters spelling out a sentence. Before a cell divides, the DNA is copied, so that each of the resulting two cells will inherit the DNA sequence. Portions of a DNA molecule that specify a single functional unit are called genes; different genes have different sequences of bases. Within cells, the long strands of DNA form condensed structures called chromosomes. The specific location of a DNA sequence within a chromosome is known as a locus. If the DNA sequence at a locus varies between individuals, the different forms of this sequence are called alleles. DNA sequences can change through mutations, producing new alleles. If a mutation occurs within a gene, the new allele may affect the trait that the gene controls, altering the phenotype of the organism. However, while this simple correspondence between an allele and a trait works in some cases, most traits are more complex and are controlled by quantitative trait loci (multiple interacting genes).

Recent findings have confirmed important examples of heritable changes that cannot be explained by changes to the sequence of nucleotides in the DNA. These phenomena are classed as epigenetic inheritance systems. DNA methylation marking chromatin, self-sustaining metabolic loops, gene silencing by RNA interference and the three-dimensional conformation of proteins (such as prions) are areas where epigenetic inheritance systems have been discovered at the organismic level. Developmental biologists suggest that complex interactions in genetic networks and communication among cells can lead to heritable variations that may underlay some of the mechanics in developmental plasticity and canalisation. Heritability may also occur at even larger scales. For example, ecological inheritance through the process of niche construction is defined by the regular and repeated activities of organisms in their environment. This generates a legacy of effects that modify and feed back into the selection regime of subsequent generations. Descendants inherit genes plus environmental characteristics generated by the ecological actions of ancestors. Other examples of heritability in evolution that are not under the direct control of genes include the inheritance of cultural traits and symbiogenesis.

Variation

An individual organism's phenotype results from both its genotype and the influence from the environment it has lived in. A substantial part of the phenotypic variation in a population is caused by genotypic variation. The modern evolutionary synthesis defines evolution as the change over time in this genetic variation. The frequency of one particular allele will become more or less prevalent relative to other forms of that gene. Variation disappears when a new allele reaches the point of fixation—when it either disappears from the population or replaces the ancestral allele entirely.

White peppered moth

Black morph in peppered moth evolution

Natural selection will only cause evolution if there is enough genetic variation in a population. Before the discovery of Mendelian genetics, one common hypothesis was blending inheritance. But with blending inheritance, genetic variance would be rapidly lost, making evolution by natural selection implausible. The Hardy–Weinberg principle provides the solution to how variation is maintained in a population with Mendelian inheritance. The frequencies of alleles (variations in a gene) will remain constant in the absence of selection, mutation, migration and genetic drift.

Variation comes from mutations in the genome, reshuffling of genes through sexual reproduction and migration between populations (gene flow). Despite the constant introduction of new variation through mutation and gene flow, most of the genome of a species is identical in all individuals of that species. However, even relatively small differences in genotype can lead to dramatic differences in phenotype: for example, chimpanzees and humans differ in only about 5% of their genomes.

Mutation

Mutations are changes in the DNA sequence of a cell's genome. When mutations occur, they may alter the product of a gene, or prevent the gene from functioning, or have no effect. Based on studies in the fly *Drosophila melanogaster*, it has been suggested that if a mutation changes a protein produced by a gene, this will probably be harmful, with about 70% of these mutations having damaging effects, and the remainder being either neutral or weakly beneficial.

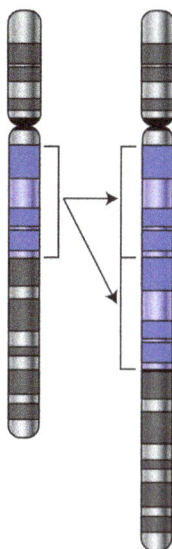

Duplication of part of a chromosome

Mutations can involve large sections of a chromosome becoming duplicated (usually by genetic recombination), which can introduce extra copies of a gene into a genome. Extra copies of genes are a major source of the raw material needed for new genes to evolve. This is important because most new genes evolve within gene families from pre-existing genes that share common ancestors. For example, the human eye uses four genes to make structures that sense light: three for colour vision and one for night vision; all four are descended from a single ancestral gene.

New genes can be generated from an ancestral gene when a duplicate copy mutates and acquires a new function. This process is easier once a gene has been duplicated because it increases the redundancy of the system; one gene in the pair can acquire a new function while the other copy continues to perform its original function. Other types of mutations can even generate entirely new genes from previously noncoding DNA.

The generation of new genes can also involve small parts of several genes being duplicated, with these fragments then recombining to form new combinations with new functions. When new genes are assembled from shuffling pre-existing parts, domains act as modules with simple independent functions, which can be mixed together to produce new combinations with new and complex functions. For example, polyketide synthases are large enzymes that make antibiotics; they contain up to one hundred independent domains that each catalyse one step in the overall process, like a step in an assembly line.

Sex and Recombination

In asexual organisms, genes are inherited together, or *linked*, as they cannot mix with genes of other organisms during reproduction. In contrast, the offspring of sexual organisms contain random mixtures of their parents' chromosomes that are produced through independent assortment.

In a related process called homologous recombination, sexual organisms exchange DNA between two matching chromosomes. Recombination and reassortment do not alter allele frequencies, but instead change which alleles are associated with each other, producing offspring with new combinations of alleles. Sex usually increases genetic variation and may increase the rate of evolution.

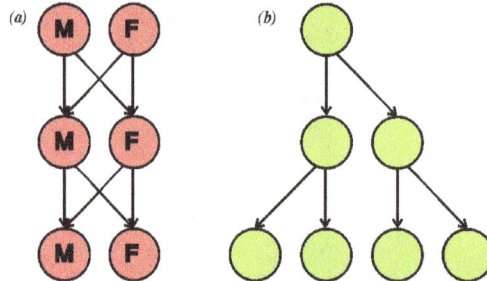

This diagram illustrates the *twofold cost of sex*. If each individual were to contribute to the same number of offspring (two), *(a)* the sexual population remains the same size each generation, where the *(b)* Asexual reproduction population doubles in size each generation.

The two-fold cost of sex was first described by John Maynard Smith. The first cost is that in sexually dimorphic species only one of the two sexes can bear young. (This cost does not apply to hermaphroditic species, like most plants and many invertebrates.) The second cost is that any individual who reproduces sexually can only pass on 50% of its genes to any individual offspring, with even less passed on as each new generation passes. Yet sexual reproduction is the more common means of reproduction among eukaryotes and multicellular organisms. The Red Queen hypothesis has been used to explain the significance of sexual reproduction as a means to enable continual evolution and adaptation in response to coevolution with other species in an ever-changing environment.

Gene Flow

Gene flow is the exchange of genes between populations and between species. It can therefore be a source of variation that is new to a population or to a species. Gene flow can be caused by the movement of individuals between separate populations of organisms, as might be caused by the movement of mice between inland and coastal populations, or the movement of pollen between heavy metal tolerant and heavy metal sensitive populations of grasses.

Gene transfer between species includes the formation of hybrid organisms and horizontal gene transfer. Horizontal gene transfer is the transfer of genetic material from one organism to another organism that is not its offspring; this is most common among bacteria. In medicine, this contributes to the spread of antibiotic resistance, as when one bacteria acquires resistance genes it can rapidly transfer them to other species. Horizontal transfer of genes from bacteria to eukaryotes such as the yeast *Saccharomyces cerevisiae* and the adzuki bean weevil *Callosobruchus chinensis* has occurred. An example of larger-scale transfers are the eukaryotic bdelloid rotifers, which have received a range of genes from bacteria, fungi and plants. Viruses can also carry DNA between organisms, allowing transfer of genes even across biological domains.

Large-scale gene transfer has also occurred between the ancestors of eukaryotic cells and bacteria, during the acquisition of chloroplasts and mitochondria. It is possible that eukaryotes themselves originated from horizontal gene transfers between bacteria and archaea.

Mechanisms

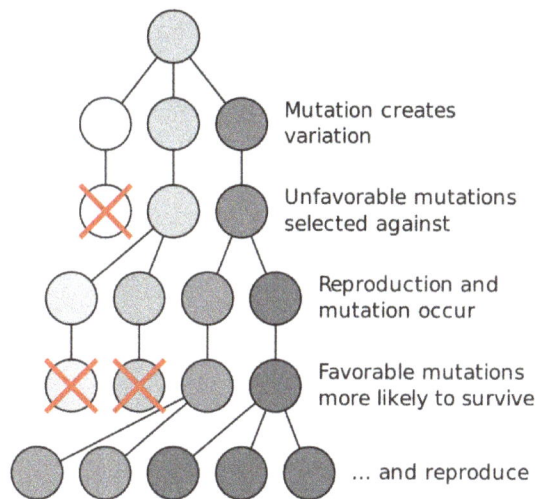

Mutation creates variation

Unfavorable mutations selected against

Reproduction and mutation occur

Favorable mutations more likely to survive

... and reproduce

Mutation followed by natural selection, results in a population with darker colouration.

From a Neo-Darwinian perspective, evolution occurs when there are changes in the frequencies of alleles within a population of interbreeding organisms. For example, the allele for black colour in a population of moths becoming more common. Mechanisms that can lead to changes in allele frequencies include natural selection, genetic drift, genetic hitchhiking, mutation and gene flow.

Natural Selection

Evolution by means of natural selection is the process by which traits that enhance survival and reproduction become more common in successive generations of a population. It has often been called a "self-evident" mechanism because it necessarily follows from three simple facts:

- Variation exists within populations of organisms with respect to morphology, physiology, and behaviour (phenotypic variation).

- Different traits confer different rates of survival and reproduction (differential fitness).

- These traits can be passed from generation to generation (heritability of fitness).

More offspring are produced than can possibly survive, and these conditions produce competition between organisms for survival and reproduction. Consequently, organisms with traits that give them an advantage over their competitors are more likely to pass on their traits to the next generation than those with traits that do not confer an advantage.

The central concept of natural selection is the evolutionary fitness of an organism. Fitness is measured by an organism's ability to survive and reproduce, which determines the size of its genetic contribution to the next generation. However, fitness is not the same as the total number of offspring: instead fitness is indicated by the proportion of subsequent generations that carry an organism's genes. For example, if an organism could survive well and reproduce rapidly, but its offspring were all too small and weak to survive, this organism would make little genetic contribution to future generations and would thus have low fitness.

If an allele increases fitness more than the other alleles of that gene, then with each generation this allele will become more common within the population. These traits are said to be "selected *for*." Examples of traits that can increase fitness are enhanced survival and increased fecundity. Conversely, the lower fitness caused by having a less beneficial or deleterious allele results in this allele becoming rarer—they are "selected *against*." Importantly, the fitness of an allele is not a fixed characteristic; if the environment changes, previously neutral or harmful traits may become beneficial and previously beneficial traits become harmful. However, even if the direction of selection does reverse in this way, traits that were lost in the past may not re-evolve in an identical form.

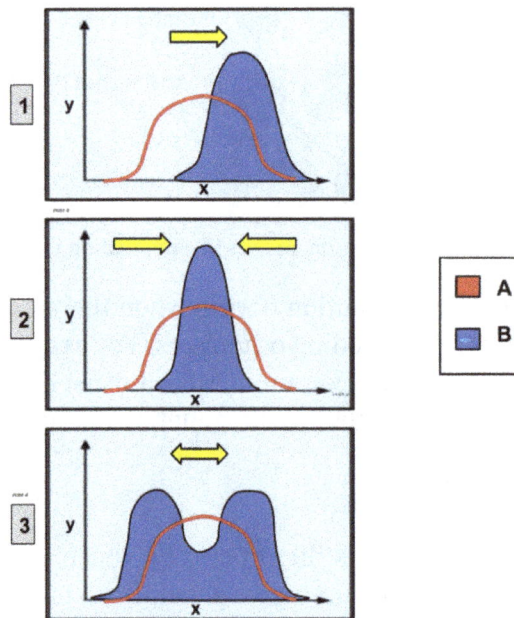

These charts depict the different types of genetic selection. On each graph, the x-axis variable is the type of phenotypic trait and the y-axis variable is the amount of organisms. Group A is the original population and Group B is the population after selection.
• Graph 1 shows directional selection, in which a single extreme phenotype is favored.
• Graph 2 depicts stabilizing selection, where the intermediate phenotype is favored over the extreme traits.
• Graph 3 shows disruptive selection, in which the extreme phenotypes are favored over the intermediate.

Natural selection within a population for a trait that can vary across a range of values, such as height, can be categorised into three different types. The first is directional selection, which is a shift in the average value of a trait over time—for example, organisms slowly getting taller. Secondly, disruptive selection is selection for extreme trait values and often results in two different values becoming most common, with selection against the average value. This would be when either short or tall organisms had an advantage, but not those of medium height. Finally, in stabilising selection there is selection against extreme trait values on both ends, which causes a decrease in variance around the average value and less diversity. This would, for example, cause organisms to slowly become all the same height.

A special case of natural selection is sexual selection, which is selection for any trait that increases mating success by increasing the attractiveness of an organism to potential mates. Traits that evolved through sexual selection are particularly prominent among males of several animal species. Although sexually favoured, traits such as cumbersome antlers, mating calls, large body size and bright colours often attract predation, which compromises the survival of individual males. This survival disadvantage is balanced by higher reproductive success in males that show these hard-to-fake, sexually selected traits.

Natural selection most generally makes nature the measure against which individuals and individual traits, are more or less likely to survive. "Nature" in this sense refers to an ecosystem, that is, a system in which organisms interact with every other element, physical as well as biological, in their local environment. Eugene Odum, a founder of ecology, defined an ecosystem as: "Any unit that includes all of the organisms...in a given area interacting with the physical environment so that a flow of energy leads to clearly defined trophic structure, biotic diversity and material cycles (ie: exchange of materials between living and nonliving parts) within the system." Each population within an ecosystem occupies a distinct niche, or position, with distinct relationships to other parts of the system. These relationships involve the life history of the organism, its position in the food chain and its geographic range. This broad understanding of nature enables scientists to delineate specific forces which, together, comprise natural selection.

Natural selection can act at different levels of organisation, such as genes, cells, individual organisms, groups of organisms and species. Selection can act at multiple levels simultaneously. An example of selection occurring below the level of the individual organism are genes called transposons, which can replicate and spread throughout a genome. Selection at a level above the individual, such as group selection, may allow the evolution of cooperation, as discussed below.

Biased Mutation

In addition to being a major source of variation, mutation may also function as a mechanism of evolution when there are different probabilities at the molecular level for different mutations to occur, a process known as mutation bias. If two genotypes, for example one with the nucleotide G and another with the nucleotide A in the same position, have the same fitness, but mutation from G to A happens more often than mutation from A to G, then genotypes with A will tend to evolve. Different insertion vs. deletion mutation biases in different taxa can lead to the evolution of different genome sizes. Developmental or mutational biases have also been observed in morphological evolution. For example, according to the phenotype-first theory of evolution, mutations can eventually cause the genetic assimilation of traits that were previously induced by the environment.

Mutation bias effects are superimposed on other processes. If selection would favor either one out of two mutations, but there is no extra advantage to having both, then the mutation that occurs the most frequently is the one that is most likely to become fixed in a population. Mutations leading to the loss of function of a gene are much more common than mutations that produce a new, fully functional gene. Most loss of function mutations are selected against. But when selection is weak, mutation bias towards loss of function can affect evolution. For example, pigments are no longer useful when animals live in the darkness of caves, and tend to be lost. This kind of loss of function can occur because of mutation bias, and/or because the function had a cost, and once the benefit of the function disappeared, natural selection leads to the loss. Loss of sporulation ability in *Bacillus subtilis* during laboratory evolution appears to have been caused by mutation bias, rather than natural selection against the cost of maintaining sporulation ability. When there is no selection for loss of function, the speed at which loss evolves depends more on the mutation rate than it does on the effective population size, indicating that it is driven more by mutation bias than by genetic drift. In parasitic organisms, mutation bias leads to selection pressures as seen in Ehrlichia. Mutations are biased towards antigenic variants in outer-membrane proteins.

Genetic Drift

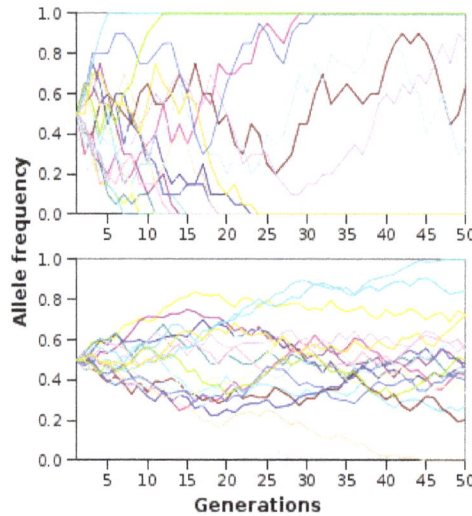

Simulation of genetic drift of 20 unlinked alleles in populations of 10 (top) and 100 (bottom). Drift to fixation is more rapid in the smaller population.

Genetic drift is the change in allele frequency from one generation to the next that occurs because alleles are subject to sampling error. As a result, when selective forces are absent or relatively weak, allele frequencies tend to "drift" upward or downward randomly (in a random walk). This drift halts when an allele eventually becomes fixed, either by disappearing from the population, or replacing the other alleles entirely. Genetic drift may therefore eliminate some alleles from a population due to chance alone. Even in the absence of selective forces, genetic drift can cause two separate populations that began with the same genetic structure to drift apart into two divergent populations with different sets of alleles.

It is usually difficult to measure the relative importance of selection and neutral processes, including drift. The comparative importance of adaptive and non-adaptive forces in driving evolutionary change is an area of current research.

The neutral theory of molecular evolution proposed that most evolutionary changes are the result of the fixation of neutral mutations by genetic drift. Hence, in this model, most genetic changes in a population are the result of constant mutation pressure and genetic drift. This form of the neutral theory is now largely abandoned, since it does not seem to fit the genetic variation seen in nature. However, a more recent and better-supported version of this model is the nearly neutral theory, where a mutation that would be effectively neutral in a small population is not necessarily neutral in a large population. Other alternative theories propose that genetic drift is dwarfed by other stochastic forces in evolution, such as genetic hitchhiking, also known as genetic draft.

The time for a neutral allele to become fixed by genetic drift depends on population size, with fixation occurring more rapidly in smaller populations. The number of individuals in a population is not critical, but instead a measure known as the effective population size. The effective population is usually smaller than the total population since it takes into account factors such as the level of inbreeding and the stage of the lifecycle in which the population is the smallest. The effective population size may not be the same for every gene in the same population.

Genetic Hitchhiking

Recombination allows alleles on the same strand of DNA to become separated. However, the rate of recombination is low (approximately two events per chromosome per generation). As a result, genes close together on a chromosome may not always be shuffled away from each other and genes that are close together tend to be inherited together, a phenomenon known as linkage. This tendency is measured by finding how often two alleles occur together on a single chromosome compared to expectations, which is called their linkage disequilibrium. A set of alleles that is usually inherited in a group is called a haplotype. This can be important when one allele in a particular haplotype is strongly beneficial: natural selection can drive a selective sweep that will also cause the other alleles in the haplotype to become more common in the population; this effect is called genetic hitchhiking or genetic draft. Genetic draft caused by the fact that some neutral genes are genetically linked to others that are under selection can be partially captured by an appropriate effective population size.

Gene Flow

Gene flow involves the exchange of genes between populations and between species. The presence or absence of gene flow fundamentally changes the course of evolution. Due to the complexity of organisms, any two completely isolated populations will eventually evolve genetic incompatibilities through neutral processes, as in the Bateson-Dobzhansky-Muller model, even if both populations remain essentially identical in terms of their adaptation to the environment.

If genetic differentiation between populations develops, gene flow between populations can introduce traits or alleles which are disadvantageous in the local population and this may lead to organisms within these populations evolving mechanisms that prevent mating with genetically distant populations, eventually resulting in the appearance of new species. Thus, exchange of genetic information between individuals is fundamentally important for the development of the biological species concept.

During the development of the modern synthesis, Sewall Wright developed his shifting balance theory, which regarded gene flow between partially isolated populations as an important aspect of adaptive evolution. However, recently there has been substantial criticism of the importance of the shifting balance theory.

Outcomes

Evolution influences every aspect of the form and behaviour of organisms. Most prominent are the specific behavioural and physical adaptations that are the outcome of natural selection. These adaptations increase fitness by aiding activities such as finding food, avoiding predators or attracting mates. Organisms can also respond to selection by cooperating with each other, usually by aiding their relatives or engaging in mutually beneficial symbiosis. In the longer term, evolution produces new species through splitting ancestral populations of organisms into new groups that cannot or will not interbreed.

These outcomes of evolution are distinguished based on time scale as macroevolution versus microevolution. Macroevolution refers to evolution that occurs at or above the level of species, in particular speciation and extinction; whereas microevolution refers to smaller evolutionary changes within a species or population, in particular shifts in gene frequency and adaptation. In general, macroevolution is regarded as the outcome of long periods of microevolution. Thus, the distinction between micro- and macroevolution is not a fundamental one—the difference is simply the time involved. However, in macroevolution, the traits of the entire species may be important. For instance, a large amount of variation among individuals allows a species to rapidly adapt to new habitats, lessening the chance of it going extinct, while a wide geographic range increases the chance of speciation, by making it more likely that part of the population will become isolated. In this sense, microevolution and macroevolution might involve selection at different levels—with microevolution acting on genes and organisms, versus macroevolutionary processes such as species selection acting on entire species and affecting their rates of speciation and extinction.

A common misconception is that evolution has goals, long-term plans, or an innate tendency for "progress," as expressed in beliefs such as orthogenesis and evolutionism; realistically however, evolution has no long-term goal and does not necessarily produce greater complexity. Although complex species have evolved, they occur as a side effect of the overall number of organisms increasing and simple forms of life still remain more common in the biosphere. For example, the overwhelming majority of species are microscopic prokaryotes, which form about half the world's biomass despite their small size, and constitute the vast majority of Earth's biodiversity. Simple organisms have therefore been the dominant form of life on Earth throughout its history and continue to be the main form of life up to the present day, with complex life only appearing more diverse because it is more noticeable. Indeed, the evolution of microorganisms is particularly important to modern evolutionary research, since their rapid reproduction allows the study of experimental evolution and the observation of evolution and adaptation in real time.

Adaptation

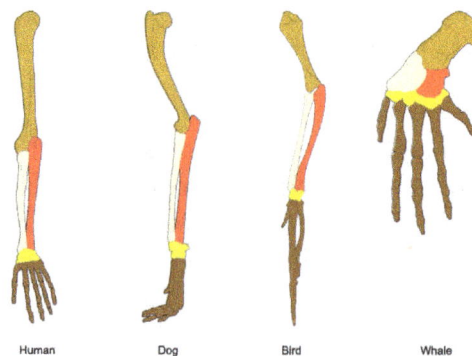

Homologous bones in the limbs of tetrapods. The bones of these animals have the same basic structure, but have been adapted for specific uses.

Adaptation is the process that makes organisms better suited to their habitat. Also, the term adaptation may refer to a trait that is important for an organism's survival. For example, the adaptation of horses' teeth to the grinding of grass. By using the term *adaptation* for the evolutionary process and *adaptive trait* for the product (the bodily part or function), the two senses of the word may be distinguished. Adaptations are produced by natural selection. The following definitions are due to Theodosius Dobzhansky:

1. *Adaptation* is the evolutionary process whereby an organism becomes better able to live in its habitat or habitats.

2. *Adaptedness* is the state of being adapted: the degree to which an organism is able to live and reproduce in a given set of habitats.

3. An *adaptive trait* is an aspect of the developmental pattern of the organism which enables or enhances the probability of that organism surviving and reproducing.

Adaptation may cause either the gain of a new feature, or the loss of an ancestral feature. An example that shows both types of change is bacterial adaptation to antibiotic selection, with genetic changes causing antibiotic resistance by both modifying the target of the drug, or increasing the activity of transporters that pump the drug out of the cell. Other striking examples are the bacteria *Escherichia coli* evolving the ability to use citric acid as a nutrient in a long-term laboratory experiment, *Flavobacterium* evolving a novel enzyme that allows these bacteria to grow on the by-products of nylon manufacturing, and the soil bacterium *Sphingobium* evolving an entirely new metabolic pathway that degrades the synthetic pesticide pentachlorophenol. An interesting but still controversial idea is that some adaptations might increase the ability of organisms to generate genetic diversity and adapt by natural selection (increasing organisms' evolvability).

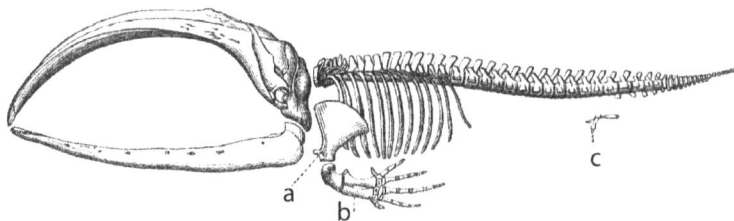

A baleen whale skeleton, *a* and *b* label flipper bones, which were adapted from front leg bones: while *c* indicates vestigial leg bones, suggesting an adaptation from land to sea.

Adaptation occurs through the gradual modification of existing structures. Consequently, structures with similar internal organisation may have different functions in related organisms. This is the result of a single ancestral structure being adapted to function in different ways. The bones within bat wings, for example, are very similar to those in mice feet and primate hands, due to the descent of all these structures from a common mammalian ancestor. However, since all living organisms are related to some extent, even organs that appear to have little or no structural similarity, such as arthropod, squid and vertebrate eyes, or the limbs and wings of arthropods and vertebrates, can depend on a common set of homologous genes that control their assembly and function; this is called deep homology.

During evolution, some structures may lose their original function and become vestigial structures. Such structures may have little or no function in a current species, yet have a clear function in ancestral species, or other closely related species. Examples include pseudogenes, the non-functional remains of eyes in blind cave-dwelling fish, wings in flightless birds, the presence of hip bones in whales and snakes, and sexual traits in organisms that reproduce via asexual reproduction. Examples of vestigial structures in humans include wisdom teeth, the coccyx, the vermiform appendix, and other behavioural vestiges such as goose bumps and primitive reflexes.

However, many traits that appear to be simple adaptations are in fact exaptations: structures originally adapted for one function, but which coincidentally became somewhat useful for some other function in the process. One example is the African lizard *Holaspis guentheri*, which developed an extremely flat head for hiding in crevices, as can be seen by looking at its near relatives. However, in this species, the head has become so flattened that it assists in gliding from tree to tree—an exaptation. Within cells, molecular machines such as the bacterial flagella and protein sorting machinery evolved by the recruitment of several pre-existing proteins that previously had different functions. Another example is the recruitment of enzymes from glycolysis and xenobiotic metabolism to serve as structural proteins called crystallins within the lenses of organisms' eyes.

An area of current investigation in evolutionary developmental biology is the developmental basis of adaptations and exaptations. This research addresses the origin and evolution of embryonic development and how modifications of development and developmental processes produce novel features. These studies have shown that evolution can alter development to produce new structures, such as embryonic bone structures that develop into the jaw in other animals instead forming part of the middle ear in mammals. It is also possible for structures that have been lost in evolution to reappear due to changes in developmental genes, such as a mutation in chickens causing embryos to grow teeth similar to those of crocodiles. It is now becoming clear that most alterations in the form of organisms are due to changes in a small set of conserved genes.

Coevolution

Common garter snake (*Thamnophis sirtalis sirtalis*) which has evolved resistance to tetrodotoxin in its amphibian prey.

Interactions between organisms can produce both conflict and cooperation. When the interaction is between pairs of species, such as a pathogen and a host, or a predator and its prey, these species can develop matched sets of adaptations. Here, the evolution of one species causes adaptations in a second species. These changes in the second species then, in turn, cause new adaptations in the first species. This cycle of selection and response is called coevolution. An example is the production of tetrodotoxin in the rough-skinned newt and the evolution of tetrodotoxin resistance in its predator, the common garter snake. In this predator-prey pair, an evolutionary arms race has produced high levels of toxin in the newt and correspondingly high levels of toxin resistance in the snake.

Cooperation

Not all co-evolved interactions between species involve conflict. Many cases of mutually beneficial interactions have evolved. For instance, an extreme cooperation exists between plants and the mycorrhizal fungi that grow on their roots and aid the plant in absorbing nutrients from the soil. This is a reciprocal relationship as the plants provide the fungi with sugars from photosynthesis. Here, the fungi actually grow inside plant cells, allowing them to exchange nutrients with their hosts, while sending signals that suppress the plant immune system.

Coalitions between organisms of the same species have also evolved. An extreme case is the eusociality found in social insects, such as bees, termites and ants, where sterile insects feed and guard the small number of organisms in a colony that are able to reproduce. On an even smaller scale, the somatic cells that make up the body of an animal limit their reproduction so they can maintain a stable organism, which then supports a small number of the animal's germ cells to produce offspring. Here, somatic cells respond to specific signals that instruct them whether to grow, remain as they are, or die. If cells ignore these signals and multiply inappropriately, their uncontrolled growth causes cancer.

Such cooperation within species may have evolved through the process of kin selection, which is where one organism acts to help raise a relative's offspring. This activity is selected for because if the *helping* individual contains alleles which promote the helping activity, it is likely that its kin will *also* contain these alleles and thus those alleles will be passed on. Other processes that may promote cooperation include group selection, where cooperation provides benefits to a group of organisms.

Speciation

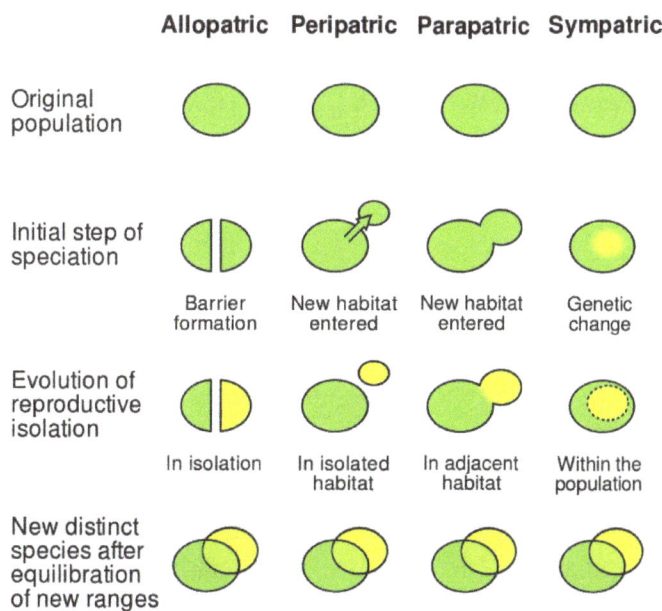

The four mechanisms of speciation

Speciation is the process where a species diverges into two or more descendant species.

There are multiple ways to define the concept of "species." The choice of definition is dependent on the particularities of the species concerned. For example, some species concepts apply more readily toward sexually reproducing organisms while others lend themselves better toward asexual organisms. Despite the diversity of various species concepts, these various concepts can be placed into one of three broad philosophical approaches: interbreeding, ecological and phylogenetic. The Biological Species Concept (BSC) is a classic example of the interbreeding approach. Defined by Ernst Mayr in 1942, the BSC states that "species are groups of actually or potentially interbreeding natural populations, which are reproductively isolated from other such groups." Despite its wide and long-term use, the BSC like others is not without controversy, for example because these concepts cannot be applied to prokaryotes, and this is called the species problem. Some researchers have attempted a unifying monistic definition of species, while others adopt a pluralistic approach and suggest that there may be different ways to logically interpret the definition of a species.

Barriers to reproduction between two diverging sexual populations are required for the populations to become new species. Gene flow may slow this process by spreading the new genetic variants also to the other populations. Depending on how far two species have diverged since their most recent common ancestor, it may still be possible for them to produce offspring, as with horses and donkeys mating to produce mules. Such hybrids are generally infertile. In this case, closely related species may regularly interbreed, but hybrids will be selected against and the species will remain distinct. However, viable hybrids are occasionally formed and these new species can either have properties intermediate between their parent species, or possess a totally new phenotype. The importance of hybridisation in producing new species of animals is unclear, although cases have been seen in many types of animals, with the gray tree frog being a particularly well-studied example.

Speciation has been observed multiple times under both controlled laboratory conditions and in nature. In sexually reproducing organisms, speciation results from reproductive isolation followed by genealogical divergence. There are four mechanisms for speciation. The most common in animals is allopatric speciation, which occurs in populations initially isolated geographically, such as by habitat fragmentation or migration. Selection under these conditions can produce very rapid changes in the appearance and behaviour of organisms. As selection and drift act independently on populations isolated from the rest of their species, separation may eventually produce organisms that cannot interbreed.

The second mechanism of speciation is peripatric speciation, which occurs when small populations of organisms become isolated in a new environment. This differs from allopatric speciation in that the isolated populations are numerically much smaller than the parental population. Here, the founder effect causes rapid speciation after an increase in inbreeding increases selection on homozygotes, leading to rapid genetic change.

The third mechanism of speciation is parapatric speciation. This is similar to peripatric speciation in that a small population enters a new habitat, but differs in that there is no physical separation between these two populations. Instead, speciation results from the evolution of mechanisms that reduce gene flow between the two populations. Generally this occurs when there has been a drastic change in the environment within the parental species' habitat. One example is the grass *Anthoxanthum odoratum*, which can undergo parapatric speciation in response to localised metal pollution from mines. Here, plants evolve that have resistance to high levels of metals in the soil.

Selection against interbreeding with the metal-sensitive parental population produced a gradual change in the flowering time of the metal-resistant plants, which eventually produced complete reproductive isolation. Selection against hybrids between the two populations may cause *reinforcement*, which is the evolution of traits that promote mating within a species, as well as character displacement, which is when two species become more distinct in appearance.

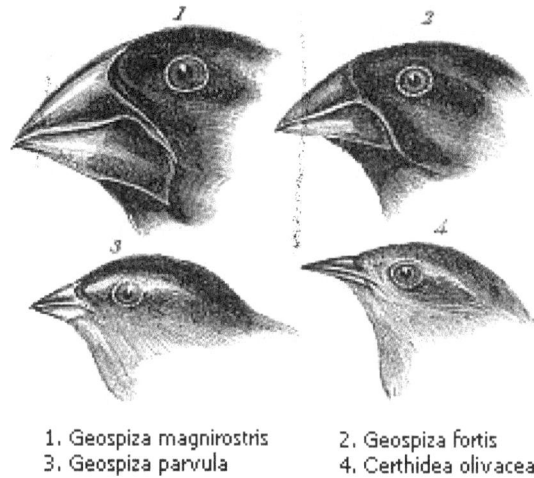

1. Geospiza magnirostris   2. Geospiza fortis
3. Geospiza parvula   4. Certhidea olivacea

Finches from Galapagos Archipelago

Geographical isolation of finches on the Galápagos Islands produced over a dozen new species.

Finally, in sympatric speciation species diverge without geographic isolation or changes in habitat. This form is rare since even a small amount of gene flow may remove genetic differences between parts of a population. Generally, sympatric speciation in animals requires the evolution of both genetic differences and non-random mating, to allow reproductive isolation to evolve.

One type of sympatric speciation involves crossbreeding of two related species to produce a new hybrid species. This is not common in animals as animal hybrids are usually sterile. This is because during meiosis the homologous chromosomes from each parent are from different species and cannot successfully pair. However, it is more common in plants because plants often double their number of chromosomes, to form polyploids. This allows the chromosomes from each parental species to form matching pairs during meiosis, since each parent's chromosomes are represented by a pair already. An example of such a speciation event is when the plant species *Arabidopsis thaliana* and *Arabidopsis arenosa* crossbred to give the new species *Arabidopsis suecica*. This happened about 20,000 years ago, and the speciation process has been repeated in the laboratory, which allows the study of the genetic mechanisms involved in this process. Indeed, chromosome doubling within a species may be a common cause of reproductive isolation, as half the doubled chromosomes will be unmatched when breeding with undoubled organisms.

Speciation events are important in the theory of punctuated equilibrium, which accounts for the pattern in the fossil record of short "bursts" of evolution interspersed with relatively long periods of stasis, where species remain relatively unchanged. In this theory, speciation and rapid evolution are linked, with natural selection and genetic drift acting most strongly on organisms undergoing speciation in novel habitats or small populations. As a result, the periods of stasis in the fossil record correspond to the parental population and the organisms undergoing speciation and rapid evolution are found in small populations or geographically restricted habitats and therefore rarely being preserved as fossils.

Extinction

Tyrannosaurus rex. Non-avian dinosaurs died out in the Cretaceous–Paleogene extinction event at the end of the Cretaceous period.

Extinction is the disappearance of an entire species. Extinction is not an unusual event, as species regularly appear through speciation and disappear through extinction. Nearly all animal and plant species that have lived on Earth are now extinct, and extinction appears to be the ultimate fate of all species. These extinctions have happened continuously throughout the history of life, although the rate of extinction spikes in occasional mass extinction events. The Cretaceous–Paleogene extinction event, during which the non-avian dinosaurs became extinct, is the most well-known, but the earlier Permian–Triassic extinction event was even more severe, with approximately 96% of all marine species driven to extinction. The Holocene extinction event is an ongoing mass extinction associated with humanity's expansion across the globe over the past few thousand years. Present-day extinction rates are 100–1000 times greater than the background rate and up to 30% of current species may be extinct by the mid 21st century. Human activities are now the primary cause of the ongoing extinction event; global warming may further accelerate it in the future.

The role of extinction in evolution is not very well understood and may depend on which type of extinction is considered. The causes of the continuous "low-level" extinction events, which form the majority of extinctions, may be the result of competition between species for limited resources (the competitive exclusion principle). If one species can out-compete another, this could produce species selection, with the fitter species surviving and the other species being driven to extinction. The intermittent mass extinctions are also important, but instead of acting as a selective force, they drastically reduce diversity in a nonspecific manner and promote bursts of rapid evolution and speciation in survivors.

Evolutionary History of Life

Origin of Life

The Earth is about 4.54 billion years old. The earliest undisputed evidence of life on Earth dates from at least 3.5 billion years ago, during the Eoarchean Era after a geological crust started to solidify following the earlier molten Hadean Eon. Microbial mat fossils have been found in 3.48 billion-year-old sandstone in Western Australia. Other early physical evidence of a biogenic substance is graphite in 3.7 billion-year-old metasedimentary rocks discovered in Western Greenland as well as "remains of biotic life" found in 4.1 billion-year-old rocks in Western Australia. According to one of the researchers, "If life arose relatively quickly on Earth ... then it could be common in the universe."

More than 99 percent of all species, amounting to over five billion species, that ever lived on Earth are estimated to be extinct. Estimates on the number of Earth's current species range from 10 million to 14 million, of which about 1.2 million have been documented and over 86 percent have not yet been described.

Highly energetic chemistry is thought to have produced a self-replicating molecule around 4 billion years ago, and half a billion years later the last common ancestor of all life existed. The current scientific consensus is that the complex biochemistry that makes up life came from simpler chemical reactions. The beginning of life may have included self-replicating molecules such as RNA and the assembly of simple cells.

Common Descent

All organisms on Earth are descended from a common ancestor or ancestral gene pool. Current species are a stage in the process of evolution, with their diversity the product of a long series of speciation and extinction events. The common descent of organisms was first deduced from four simple facts about organisms: First, they have geographic distributions that cannot be explained by local adaptation. Second, the diversity of life is not a set of completely unique organisms, but organisms that share morphological similarities. Third, vestigial traits with no clear purpose resemble functional ancestral traits and finally, that organisms can be classified using these similarities into a hierarchy of nested groups—similar to a family tree. However, modern research has suggested that, due to horizontal gene transfer, this "tree of life" may be more complicated than a simple branching tree since some genes have spread independently between distantly related species.

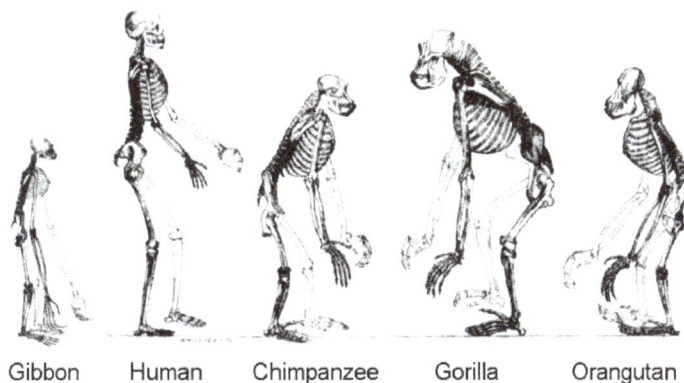

Gibbon   Human   Chimpanzee   Gorilla   Orangutan

The hominoids are descendants of a common ancestor.

Past species have also left records of their evolutionary history. Fossils, along with the comparative anatomy of present-day organisms, constitute the morphological, or anatomical, record. By comparing the anatomies of both modern and extinct species, paleontologists can infer the lineages of those species. However, this approach is most successful for organisms that had hard body parts, such as shells, bones or teeth. Further, as prokaryotes such as bacteria and archaea share a limited set of common morphologies, their fossils do not provide information on their ancestry.

More recently, evidence for common descent has come from the study of biochemical similarities between organisms. For example, all living cells use the same basic set of nucleotides and amino acids. The development of molecular genetics has revealed the record of evolution left in organisms' genomes: dating when species diverged through the molecular clock produced by mutations. For example, these DNA sequence comparisons have revealed that humans and chimpanzees share 98% of their genomes and analysing the few areas where they differ helps shed light on when the common ancestor of these species existed.

Evolution of Life

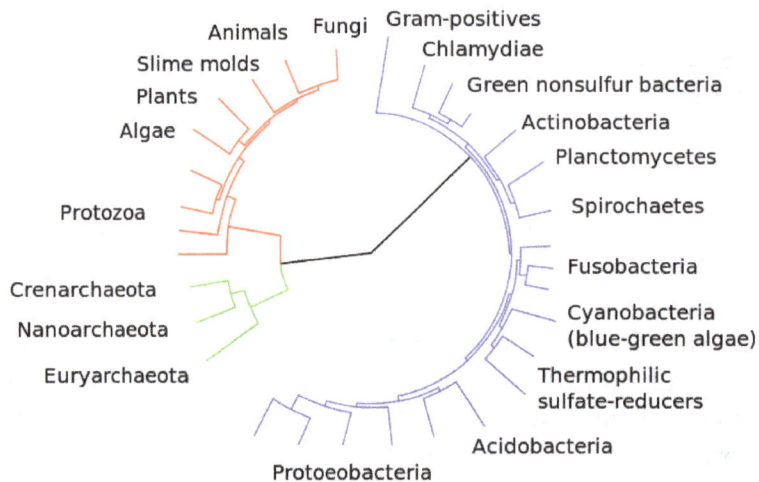

Evolutionary tree showing the divergence of modern species from their common ancestor in the centre. The three domains are coloured, with bacteria blue, archaea green and eukaryotes red.

Prokaryotes inhabited the Earth from approximately 3–4 billion years ago. No obvious changes in morphology or cellular organisation occurred in these organisms over the next few billion years. The eukaryotic cells emerged between 1.6–2.7 billion years ago. The next major change in cell structure came when bacteria were engulfed by eukaryotic cells, in a cooperative association called endosymbiosis. The engulfed bacteria and the host cell then underwent coevolution, with the bacteria evolving into either mitochondria or hydrogenosomes. Another engulfment of cyanobacterial-like organisms led to the formation of chloroplasts in algae and plants.

The history of life was that of the unicellular eukaryotes, prokaryotes and archaea until about 610 million years ago when multicellular organisms began to appear in the oceans in the Ediacaran period. The evolution of multicellularity occurred in multiple independent events, in organisms as diverse as sponges, brown algae, cyanobacteria, slime moulds and myxobacteria. In January 2016, scientists reported that, about 800 million years ago, a minor genetic change in a single molecule called GK-PID may have allowed organisms to go from a single cell organism to one of many cells.

Soon after the emergence of these first multicellular organisms, a remarkable amount of biological diversity appeared over approximately 10 million years, in an event called the Cambrian explosion.

Here, the majority of types of modern animals appeared in the fossil record, as well as unique lineages that subsequently became extinct. Various triggers for the Cambrian explosion have been proposed, including the accumulation of oxygen in the atmosphere from photosynthesis.

About 500 million years ago, plants and fungi colonised the land and were soon followed by arthropods and other animals. Insects were particularly successful and even today make up the majority of animal species. Amphibians first appeared around 364 million years ago, followed by early amniotes and birds around 155 million years ago (both from "reptile"-like lineages), mammals around 129 million years ago, homininae around 10 million years ago and modern humans around 250,000 years ago. However, despite the evolution of these large animals, smaller organisms similar to the types that evolved early in this process continue to be highly successful and dominate the Earth, with the majority of both biomass and species being prokaryotes.

Applications

Concepts and models used in evolutionary biology, such as natural selection, have many applications.

Artificial selection is the intentional selection of traits in a population of organisms. This has been used for thousands of years in the domestication of plants and animals. More recently, such selection has become a vital part of genetic engineering, with selectable markers such as antibiotic resistance genes being used to manipulate DNA. Proteins with valuable properties have evolved by repeated rounds of mutation and selection (for example modified enzymes and new antibodies) in a process called directed evolution.

Understanding the changes that have occurred during an organism's evolution can reveal the genes needed to construct parts of the body, genes which may be involved in human genetic disorders. For example, the Mexican tetra is an albino cavefish that lost its eyesight during evolution. Breeding together different populations of this blind fish produced some offspring with functional eyes, since different mutations had occurred in the isolated populations that had evolved in different caves. This helped identify genes required for vision and pigmentation.

Many human diseases are not static phenomena, but capable of evolution. Viruses, bacteria, fungi and cancers evolve to be resistant to host immune defences, as well as pharmaceutical drugs. These same problems occur in agriculture with pesticide and herbicide resistance. It is possible that we are facing the end of the effective life of most of available antibiotics and predicting the evolution and evolvability of our pathogens and devising strategies to slow or circumvent it is requiring deeper knowledge of the complex forces driving evolution at the molecular level.

In computer science, simulations of evolution using evolutionary algorithms and artificial life started in the 1960s and were extended with simulation of artificial selection. Artificial evolution became a widely recognised optimisation method as a result of the work of Ingo Rechenberg in the 1960s. He used evolution strategies to solve complex engineering problems. Genetic algorithms in particular became popular through the writing of John Henry Holland. Practical applications also include automatic evolution of computer programmes. Evolutionary algorithms are now used to solve multi-dimensional problems more efficiently than software produced by human designers and also to optimise the design of systems.

Social and Cultural Responses

In the 19th century, particularly after the publication of *On the Origin of Species* in 1859, the idea that life had evolved was an active source of academic debate centred on the philosophical, social and religious implications of evolution. Today, the modern evolutionary synthesis is accepted by a vast majority of scientists. However, evolution remains a contentious concept for some theists.

As evolution became widely accepted in the 1870s, caricatures of Charles Darwin with an ape or monkey body symbolised evolution.

While various religions and denominations have reconciled their beliefs with evolution through concepts such as theistic evolution, there are creationists who believe that evolution is contradicted by the creation myths found in their religions and who raise various objections to evolution. As had been demonstrated by responses to the publication of *Vestiges of the Natural History of Creation* in 1844, the most controversial aspect of evolutionary biology is the implication of human evolution that humans share common ancestry with apes and that the mental and moral faculties of humanity have the same types of natural causes as other inherited traits in animals. In some countries, notably the United States, these tensions between science and religion have fuelled the current creation–evolution controversy, a religious conflict focusing on politics and public education. While other scientific fields such as cosmology and Earth science also conflict with literal interpretations of many religious texts, evolutionary biology experiences significantly more opposition from religious literalists.

The teaching of evolution in American secondary school biology classes was uncommon in most of the first half of the 20th century. The Scopes Trial decision of 1925 caused the subject to become very rare in American secondary biology textbooks for a generation, but it was gradually re-introduced later and became legally protected with the 1968 *Epperson v. Arkansas* decision. Since then, the competing religious belief of creationism was legally disallowed in secondary school curricula in various decisions in the 1970s and 1980s, but it returned in pseudoscientific form as intelligent design (ID), to be excluded once again in the 2005 *Kitzmiller v. Dover Area School District* case.

Molecular Evolution

Molecular evolution is the process of change in the sequence composition of cellular molecules such as DNA, RNA, and proteins across generations. The field of molecular evolution uses principles of evolutionary biology and population genetics to explain patterns in these changes. Major topics in molecular evolution concern the rates and impacts of single nucleotide changes, neutral evolution vs. natural selection, origins of new genes, the genetic nature of complex traits, the genetic basis of speciation, evolution of development, and ways that evolutionary forces influence genomic and phenotypic changes.

Forces in Molecular Evolution

The content and structure of a genome is the product of the molecular and population genetic forces which act upon that genome. Novel genetic variants will arise through mutation and will spread and be maintained in populations due to genetic drift or natural selection.

Mutation

This hedgehog has no pigmentation due to a genetic mutation.

Mutations are permanent, transmissible changes to the genetic material (DNA or RNA) of a cell or virus. Mutations result from errors in DNA replication during cell division and by exposure to radiation, chemicals, and other environmental stressors, or viruses and transposable elements. Most mutations that occur are single nucleotide polymorphisms which modify single bases of the DNA sequence, resulting in point mutations. Other types of mutations modify larger segments of DNA and can cause duplications, insertions, deletions, inversions, and translocations.

Most organisms display a strong bias in the types of mutations that occur with strong influence in GC-content. Transitions (A ↔ G or C ↔ T) are more common than transversions (purine (adenine or guanine)) ↔ pyrimidine (cytosine or thymine, or in RNA, uracil)) and are less likely to alter amino acid sequences of proteins.

Mutations are stochastic and typically occur randomly across genes. Mutation rates for single nucleotide sites for most organisms are very low, roughly 10^{-9} to 10^{-8} per site per generation, though some viruses have higher mutation rates on the order of 10^{-6} per site per generation. Among these mutations, some will be neutral or beneficial and will remain in the genome unless lost via genetic drift, and others will be detrimental and will be eliminated from the genome by natural selection.

Because mutations are extremely rare, they accumulate very slowly across generations. While the number of mutations which appears in any single generation may vary, over very long time periods they will appear to accumulate at a regular pace. Using the mutation rate per generation and the number of nucleotide differences between two sequences, divergence times can be estimated effectively via the molecular clock.

Recombination

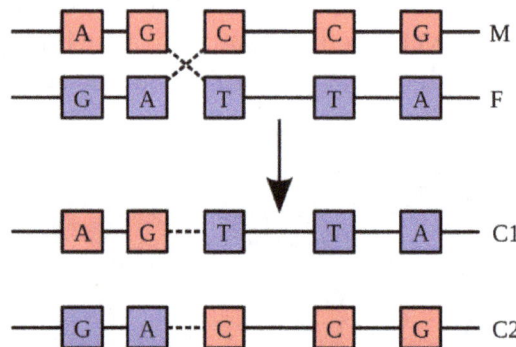

Recombination involves the breakage and rejoining of two chromosomes (M and F) to produce two re-arranged chromosomes (C1 and C2).

Recombination is a process that results in genetic exchange between chromosomes or chromosomal regions. Recombination counteracts physical linkage between adjacent genes, thereby reducing genetic hitchhiking. The resulting independent inheritance of genes results in more efficient selection, meaning that regions with higher recombination will harbor fewer detrimental mutations, more selectively favored variants, and fewer errors in replication and repair. Recombination can also generate particular types of mutations if chromosomes are misaligned.

Gene Conversion

Gene conversion is a type of recombination that is the product of DNA repair where nucleotide damage is corrected using an homologous genomic region as a template. Damaged bases are first excised, the damaged strand is then aligned with an undamaged homolog, and DNA synthesis repairs the excised region using the undamaged strand as a guide. Gene conversion is often responsible for homogenizing sequences of duplicate genes over long time periods, reducing nucleotide divergence.

Genetic Drift

Genetic drift is the change of allele frequencies from one generation to the next due to stochastic effects of random sampling in finite populations. Some existing variants have no effect on fitness and may increase or decrease in frequency simply due to chance. "Nearly neutral" variants whose selection coefficient is close to a threshold value of 1 / the effective population size will also be affected by chance as well as by selection and mutation. Many genomic features have been ascribed to accumulation of nearly neutral detrimental mutations as a result of small effective population sizes. With a smaller effective population size, a larger variety of mutations will behave as if they are neutral due to inefficiency of selection.

Selection

Selection occurs when organisms with greater fitness, i.e. greater ability to survive or reproduce, are favored in subsequent generations, thereby increasing the instance of underlying genetic variants in a population. Selection can be the product of natural selection, artificial selection, or sexual selection. Natural selection is any selective process that occurs due to the fitness of an organism to its environment. In contrast sexual selection is a product of mate choice and can favor the spread of genetic variants which act counter to natural selection but increase desirability to the opposite sex or increase mating success. Artificial selection, also known as selective breeding, is imposed by an outside entity, typically humans, in order to increase the frequency of desired traits.

The principles of population genetics apply similarly to all types of selection, though in fact each may produce distinct effects due to clustering of genes with different functions in different parts of the genome, or due to different properties of genes in particular functional classes. For instance, sexual selection could be more likely to affect molecular evolution of the sex chromosomes due to clustering of sex specific genes on the X,Y,Z or W.

Selection can operate at the gene level at the expense of organismal fitness, resulting in a selective advantage for selfish genetic elements in spite of a host cost. Examples of such selfish elements include transposable elements, meiotic drivers, killer X chromosomes, selfish mitochondria, and self-propagating introns.

Genome Architecture

Genome Size

Genome size is influenced by the amount of repetitive DNA as well as number of genes in an organism. The C-value paradox refers to the lack of correlation between organism 'complexity' and genome size. Explanations for the so-called paradox are two-fold. First, repetitive genetic elements can comprise large portions of the genome for many organisms, thereby inflating DNA content of the haploid genome. Secondly, the number of genes is not necessarily indicative of the number of developmental stages or tissue types in an organism. An organism with few developmental stages or tissue types may have large numbers of genes that influence non-developmental phenotypes, inflating gene content relative to developmental gene families.

Neutral explanations for genome size suggest that when population sizes are small, many mutations become nearly neutral. Hence, in small populations repetitive content and other 'junk' DNA can accumulate without placing the organism at a competitive disadvantage. There is little evidence to suggest that genome size is under strong widespread selection in multicellular eukaryotes. Genome size, independent of gene content, correlates poorly with most physiological traits and many eukaryotes, including mammals, harbor very large amounts of repetitive DNA.

However, birds likely have experienced strong selection for reduced genome size, in response to changing energetic needs for flight. Birds, unlike humans, produce nucleated red blood cells, and larger nuclei lead to lower levels of oxygen transport. Bird metabolism is far higher than that of mammals, due largely to flight, and oxygen needs are high. Hence, most birds have small, compact genomes with few repetitive elements. Indirect evidence suggests that non-avian theropod dinosaur ancestors of modern birds also had reduced genome sizes, consistent with endothermy and high energetic needs for running speed. Many bacteria have also experienced selection for small genome size, as time of replication and energy consumption are so tightly correlated with fitness.

Repetitive Elements

Transposable elements are self-replicating, selfish genetic elements which are capable of proliferating within host genomes. Many transposable elements are related to viruses, and share several proteins in common.

DNA transposons are cut and paste transposable elements which excise DNA and move it to alternate sections of the genome.

non-LTR retrotransposons

LTR retrotransposons

Helitrons

Alu elements comprise over 10% of the human genome. They are short non-autonomous repeat sequences.

Chromosome Number and Organization

The number of chromosomes in an organism's genome also does not necessarily correlate with the amount of DNA in its genome. The ant *Myrmecia pilosula* has only a single pair of chromosomes whereas the Adders-tongue fern *Ophioglossum reticulatum* has up to 1260 chromosomes. Cilliate genomes house each gene in individual chromosomes, resulting in a genome which is not physically linked. Reduced linkage through creation of additional chromosomes should effectively increase the efficiency of selection.

Changes in chromosome number can play a key role in speciation, as differing chromosome numbers can serve as a barrier to reproduction in hybrids. Human chromosome 2 was created from a fusion of two chimpanzee chromosomes and still contains central telomeres as well as a vestigial second centromere. Polyploidy, especially allopolyploidy, which occurs often in plants, can also result in reproductive incompatibilities with parental species. *Agrodiatus* blue butterflies have diverse chromosome numbers ranging from n=10 to n=134 and additionally have one of the highest rates of speciation identified to date.

Gene content and Distribution

Different organisms house different numbers of genes within their genomes as well as different patterns in the distribution of genes throughout the genome. Some organisms, such as most bacteria, *Drosophila*, and *Arabidopsis* have particularly compact genomes with little repetitive content or non-coding DNA. Other organisms, like mammals or maize, have large amounts of repetitive DNA, long introns, and substantial spacing between different genes. The content and distribution of genes within the genome can influence the rate at which certain types of mutations occur and can influence the subsequent evolution of different species. Genes with longer introns are more likely to recombine due to increased physical distance over the coding sequence. As such, long introns may facilitate ectopic recombination, and result in higher rates of new gene formation.

Organelles

In addition to the nuclear genome, endosymbiont organelles contain their own genetic material typically as circular plasmids. Mitochondrial and chloroplast DNA varies across taxa, but membrane-bound proteins, especially electron transport chain constituents are most often encoded in the organelle. Chloroplasts and mitochondria are maternally inherited in most species, as the organelles must pass through the egg. In a rare departure, some species of mussels are known to inherit mitochondria from father to son.

Origins of New Genes

New genes arise from several different genetic mechanisms including gene duplication, de novo origination, retrotransposition, chimeric gene formation, recruitment of non-coding sequence, and gene truncation.

Gene duplication initially leads to redundancy. However, duplicated gene sequences can mutate to develop new functions or specialize so that the new gene performs a subset of the original ancestral functions. In addition to duplicating whole genes, sometimes only a domain or part of a protein is duplicated so that the resulting gene is an elongated version of the parental gene.

Retrotransposition creates new genes by copying mRNA to DNA and inserting it into the genome. Retrogenes often insert into new genomic locations, and often develop new expression patterns and functions.

Chimeric genes form when duplication, deletion, or incomplete retrotransposition combine portions of two different coding sequences to produce a novel gene sequence. Chimeras often cause regulatory changes and can shuffle protein domains to produce novel adaptive functions.

De novo origin. Novel genes can also arise from previously non-coding DNA. For instance, Levine and colleagues reported the origin of five new genes in the D. melanogaster genome from noncoding DNA. Similar de novo origin of genes has been also shown in other organisms such as yeast, rice and humans. De novo genes may evolve from transcripts that are already expressed at low levels. Mutation of a stop codon to a regular codon or a frameshift may cause an extended protein that includes a previously non-coding sequence.

Molecular Phylogenetics

Molecular systematics is the product of the traditional fields of systematics and molecular genetics. It uses DNA, RNA, or protein sequences to resolve questions in systematics, i.e. about their correct scientific classification or taxonomy from the point of view of evolutionary biology.

Molecular systematics has been made possible by the availability of techniques for DNA sequencing, which allow the determination of the exact sequence of nucleotides or *bases* in either DNA or RNA. At present it is still a long and expensive process to sequence the entire genome of an organism, and this has been done for only a few species. However, it is quite feasible to determine the sequence of a defined area of a particular chromosome. Typical molecular systematic analyses require the sequencing of around 1000 base pairs.

The Driving Forces of Evolution

Depending on the relative importance assigned to the various forces of evolution, three perspectives provide evolutionary explanations for molecular evolution.

Selectionist hypotheses argue that selection is the driving force of molecular evolution. While acknowledging that many mutations are neutral, selectionists attribute changes in the frequencies of neutral alleles to linkage disequilibrium with other loci that are under selection, rather than to random genetic drift. Biases in codon usage are usually explained with reference to the ability of even weak selection to shape molecular evolution.

Neutralist hypotheses emphasize the importance of mutation, purifying selection, and random genetic drift. The introduction of the neutral theory by Kimura, quickly followed by King and Jukes' own findings, led to a fierce debate about the relevance of neodarwinism at the molecular level. The Neutral theory of molecular evolution proposes that most mutations in DNA are at locations not important to function or fitness. These neutral changes drift towards fixation within a population. Positive changes will be very rare, and so will not greatly contribute to DNA polymorphisms. Deleterious mutations will also not contribute very much to DNA diversity because they negatively affect fitness and so will not stay in the gene pool for long. This theory provides a framework for the molecular clock. The fate of neutral mutations are governed by genetic drift, and contribute to both nucleotide polymorphism and fixed differences between species.

In the strictest sense, the neutral theory is not accurate. Subtle changes in DNA very often have effects, but sometimes these effects are too small for natural selection to act on. Even synonymous mutations are not necessarily neutral because there is not a uniform amount of each codon. The nearly neutral theory expanded the neutralist perspective, suggesting that several mutations are nearly neutral, which means both random drift and natural selection is relevant to their dynamics. The main difference between the neutral theory and nearly neutral theory is that the latter focuses on weak selection, not strictly neutral.

Mutationists hypotheses emphasize random drift and biases in mutation patterns. Sueoka was the first to propose a modern mutationist view. He proposed that the variation in GC content was not the result of positive selection, but a consequence of the GC mutational pressure.

Protein Evolution

Protein evolution describes the changes over time in protein shape, function, and composition. Through quantitative analysis and experimentation, scientists have strived to understand the rate and causes of protein evolution. Using the amino acid sequences of hemoglobin and cytochrome c from multiple species, scientists were able to derive estimations of protein evolution rates. What they found was that the rates were not the same among proteins. Each protein has its own rate, and that rate is constant across phylogenies (i.e., hemoglobin does not evolve at the same rate as cytochrome c, but hemoglobins from humans, mice, etc. do have comparable rates of evolution.). Not all regions within a protein mutate at the same rate; functionally important areas mutate more slowly and amino acid substitutions involving similar amino acids occurs more often than dissimilar substitutions. Overall, the level of polymorphisms in proteins seems to be fairly constant. Several species (including humans, fruit flies, and mice) have similar levels of protein polymorphism.

Lipase Sequence Homology in Different Human Tissues

Query hit (click to show/hide alignment)	Target hit	Target len	Identity	Tot. score	E-value
Lipoprotein lipase (LPL) [NX_P06858-1]		475aa	100%	2570	0.0e+00
Endothelial lipase (LIPG) [NX_Q9Y5X9-1]		500aa	45%	1158	1.4e-126
Hepatic triacylglycerol lipase (LIPC) [NX_P11150-1]		499aa	43%	1037	1.5e-112
Endothelial lipase (LIPG) [NX_Q9Y5X9-2]		354aa	34%	935	1.1e-100
Pancreatic triacylglycerol lipase (PNLIP) [NX_P16233-1]		465aa	27%	503	1.2e-50
Inactive pancreatic lipase-related protein 1 (PNLIPRP1) [NX_P54315-1]		467aa	27%	497	6.4e-50
Pancreatic lipase-related protein 2 (PNLIPRP2) [NX_P54317-1]		469aa	25%	459	2.0e-45
Pancreatic lipase-related protein 3 (PNLIPRP3) [NX_Q17RR3-1]		467aa	24%	430	4.4e-42
Lipase member H (LIPH) [NX_Q8WWY8-1]		451aa	22%	423	2.9e-41
Lipase member I (LIPI) [NX_Q6XZB0-1]		460aa	21%	412	5.7e-40
Lipase member I (LIPI) [NX_Q6XZB0-2]		481aa	21%	411	6.3e-40
Lipase member I (LIPI) [NX_Q6XZB0-6]		375aa	22%	408	1.4e-39
Lipase member I (LIPI) [NX_Q6XZB0-3]		454aa	20%	405	3.1e-39

This chart compares the sequence identity of different lipase proteins throughout the human body. It demonstrates how proteins evolve, keeping some regions conserved while others change dramatically.

Relation to Nucleic Acid Evolution

Protein evolution is inescapably tied to changes and selection of DNA polymorphisms and mutations because protein sequences change in response to alterations in the DNA sequence. Amino acid sequences and nucleic acid sequences do not mutate at the same rate. Due to the degenerate nature of DNA, bases can change without affecting the amino acid sequence. For example, there are six codons that code for leucine. Thus, despite the difference in mutation rates, it is essential to incorporate nucleic acid evolution into the discussion of protein evolution. At the end of the 1960s, two groups of scientists—Kimura (1968) and King and Jukes (1969)-- independently proposed that a majority of the evolutionary changes observed in proteins were neutral. Since then, the neutral theory has been expanded upon and debated.

Discordance with Morphological Evolution

There are sometimes discordances between molecular and morphological evolution, which are reflected in molecular and morphological systematic studies, especially of bacteria, archaea and eukaryotic microbes. These discordances can be categorized as two types: (i) one morphology, multiple lineages (e.g. morphological convergence, cryptic species) and (ii) one lineage, multiple morphologies (e.g. phenotypic plasticity, multiple life-cycle stages). Neutral evolution possibly could explain the incongruences in some cases.

Journals and societies

The Society for Molecular Biology and Evolution publishes the journals "Molecular Biology and Evolution" and "Genome Biology and Evolution" and holds an annual international meeting. Other journals dedicated to molecular evolution include *Journal of Molecular Evolution* and *Molecular Phylogenetics and Evolution*. Research in molecular evolution is also published in journals of genetics, molecular biology, genomics, systematics, and evolutionary biology.

Experimental Evolution

Experimental evolution is the use of experiments or controlled field manipulations to explore evolutionary dynamics. Evolution may be observed in the laboratory as populations adapt to new environmental conditions and/or change by such stochastic processes as random genetic drift. With modern molecular tools, it is possible to pinpoint the mutations that selection acts upon, what brought about the adaptations, and to find out how exactly these mutations work. Because of the large number of generations required for adaptation to occur, evolution experiments are typically carried out with microorganisms such as bacteria, yeast or viruses, or other organisms with rapid generation times. However, laboratory studies with foxes and with rodents have shown that notable adaptations can occur within as few as 10-20 generations and experiments with wild guppies have observed adaptations within comparable numbers of generations. More recently, using experimental evolution followed by whole genome pooled sequencing, an approach known as Evolve and Resequence (E&R) is becoming popular in fruit flies.

History

Domestication and Breeding

This Chihuahua mix and Great Dane show the wide range of dog breed sizes created using artificial selection.

Unwittingly, humans have carried out evolution experiments for as long as they have been domesticating plants and animals. Selective breeding of plants and animals has led to varieties that differ dramatically from their original wild-type ancestors. Examples are the cabbage varieties, maize, or the large number of different dog breeds. The power of human breeding to create varieties with extreme differences from a single species was already recognized by Charles Darwin. In fact, he started out his book *The Origin of Species* with a chapter on variation in domestic animals. In this chapter, Darwin discussed in particular the pigeon.

Altogether at least a score of pigeons might be chosen, which if shown to an ornithologist, and he were told that they were wild birds, would certainly, I think, be ranked by him as well-defined species. Moreover, I do not believe that any ornithologist would place the English carrier, the short-faced tumbler, the runt, the barb, pouter, and fantail in the same genus; more especially as in each of these breeds several truly-inherited sub-breeds, or species as he might have called them, could be shown him. (...) I am fully convinced that the common opinion of naturalists is correct, namely, that all have descended from the rock-pigeon (*Columba livia*), including under this term several geographical races or sub-species, which differ from each other in the most trifling respects.

— *Charles Darwin, The Origin of Species*

Early Experimental Evolution

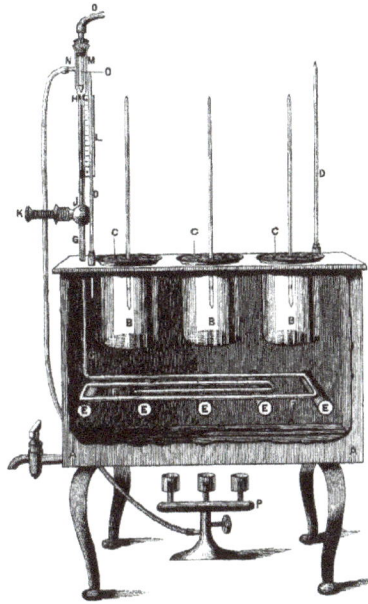

Drawing of the incubator used by Dallinger in his evolution experiments.

One of the first to carry out a controlled evolution experiment was William Dallinger. In the late 19th century, he cultivated small unicellular organisms in a custom-built incubator over a time period of seven years (1880–1886). Dallinger slowly increased the temperature of the incubator from an initial 60 °F up to 158 °F. The early cultures had shown clear signs of distress at a temperature of 73 °F, and were certainly not capable of surviving at 158 °F. The organisms Dallinger had in his incubator at the end of the experiment, on the other hand, were perfectly fine at 158 °F. However, these organisms would no longer grow at the initial 60 °F. Dallinger concluded that he had found evidence for Darwinian adaptation in his incubator, and that the organisms had adapted to live in a high-temperature environment. Unfortunately, Dallinger's incubator was accidentally destroyed in 1886, and Dallinger could not continue this line of research.

From the 1880s to 1980, experimental evolution was intermittently practiced by a variety of evolutionary biologists, including the highly influential Theodosius Dobzhansky. Like other experimental research in evolutionary biology during this period, much of this work lacked extensive replication and was carried out only for relatively short periods of evolutionary time.

Modern Experimental Evolution

Experimental evolution has been used in various formats to understand underlying evolutionary processes in a controlled system. Experimental evolution has been performed on multicellular and unicellular eukaryotes, prokaryotes, viruses. Similar works have also been performed by directed evolution of individual enzyme, ribozyme and replicator genes.

Fruit Flies

One of the first of a new wave of experiments using this strategy was the laboratory "evolutionary radiation" of *Drosophila melanogaster* populations that Michael R. Rose started in February, 1980. This system started with ten populations, five cultured at later ages, and five cultured at early ages. Since then more than 200 different populations have been created in this laboratory radiation, with selection targeting multiple characters. Some of these highly differentiated populations have also been selected "backward" or "in reverse," by returning experimental populations to their ancestral culture regime. Hundreds of people have worked with these populations over the better part of three decades. Much of this work is summarized in the papers collected in the book *Methuselah Flies*, listed below.

The early experiments in flies were limited to studying phenotypes but the molecular mechanisms, i.e., changes in DNA that facilitated such changes, could not be identified. This changed with genomics technology. Subsequently, Thomas Turner coined the term Evolve and Resequence (E&R) and several studies used E&R approach with mixed success (reviewed in and). One of the more interesting experimental evolution studies was conduced by Gabriel Haddad's group at UC San Diego, where Haddad and colleagues evolved flies to adapt to low oxygen environments, also known as hypoxia. After 200 generations, they used E&R approach to identify genomic regions that were selected by natural selection in the hypoxia adapted flies. More recent experiments are have started following up E&R predictions with RNAseq and genetic crosses. Such efforts in combining E&R with experimental validations should be powerful in identifying genes that regulate adaptation in flies.

Bacteria

Lenski's *E. Coli* Experiment

On February 15, 1988, Richard Lenski started a long-term evolution experiment with the bacterium *E. coli*. When one of the flasks suddenly developed the ability to metabolize citrate from the growth medium aerobically and showed greatly increased growth, this provided a dramatic observation of evolution in action. The experiment continues to this day, and is now the longest-running controlled evolution experiment ever undertaken. Since the inception of the experiment, the bacteria have grown for more than 60,000 generations. Lenski and colleagues regularly publish updates on the status of the experiments.

Hyperswarming in *P. Aeruginosa*

In 2013, Joao Xavier published on an experiment in which *Pseudomonas aeruginosa*, when subjected to repeated rounds of conditions in which it needed to swarm to acquire food, developed the ability to "hyperswarm" at speeds 25% faster than baseline organisms, by developing multiple flagella, whereas the baseline organism has a single flagellum. Especially notable was that this development was astonishingly repeatable.

Laboratory House Mice

Mouse from the Garland selection experiment with attached wheel (1.1 m circumference) and its photocell-based counter.

In 1998, Theodore Garland, Jr. and colleagues started a long-term experiment that involves selective breeding for high voluntary activity levels on running wheels. This experiment also continues to this day (> 65 generations). Mice from the four replicate "High Runner" lines evolved to run almost three times as many running-wheel revolutions per day compared with the four unselected control lines of mice, mainly by running faster than the control mice rather than running for more minutes/day.

Female mouse with her litter, from the Garland selection experiment.

The HR mice exhibit an elevated maximal aerobic capacity when tested on a motorized treadmill. They also exhibit alterations in motivation and the reward system of the brain. Pharmacological studies point to alterations in dopamine function and the endocannabinoid system. The High Runner lines have been proposed as a model to study human attention-deficit hyperactivity disorder (ADHD), and administration of Ritalin reduces their wheel running approximately to the levels of Control mice. Click here for a mouse wheel running video.

Other Examples

Stickleback fish have both marine and freshwater species, the freshwater species evolving since the last ice age. Fresh water species can survive colder temperatures. Scientists tested to see if they could reproduce this evolution of cold-tolerance by keeping marine sticklebacks in cold freshwater. It took the marine sticklebacks only three generations to evolve to match the 2.5 degree Celsius improvement in cold-tolerance found in wild freshwater sticklebacks.

Experimental Evolution for Teaching

Because of their rapid generation times microbes offer an opportunity to study microevolution in the classroom. A number of exercises involving bacteria and yeast teach concepts ranging from the evolution of resistance to the evolution of multicellularity. With the advent of next-generation sequencing technology it has become possible for students to conduct an evolutionary experiment, sequence the evolved genomes, and to analyze and interpret the results.

Genetic Drift

Genetic drift (also known as allelic drift or the Sewall Wright effect after biologist Sewall Wright) is the change in the frequency of a gene variant (allele) in a population due to random sampling of organisms. The alleles in the offspring are a sample of those in the parents, and chance has a role in determining whether a given individual survives and reproduces. A population's allele frequency is the fraction of the copies of one gene that share a particular form. Genetic drift may cause gene variants to disappear completely and thereby reduce genetic variation.

Biologist and statistician Ronald Fisher

When there are few copies of an allele, the effect of genetic drift is larger, and when there are many copies the effect is smaller. In the early 20th century, vigorous debates occurred over the relative importance of natural selection versus neutral processes, including genetic drift. Ronald Fisher, who explained natural selection using Mendelian genetics, held the view that genetic drift plays at the most a minor role in evolution, and this remained the dominant view for several decades. In 1968, population geneticist Motoo Kimura rekindled the debate with his neutral theory of molecular evolution, which claims that most instances where a genetic change spreads across a population (although not necessarily changes in phenotypes) are caused by genetic drift acting on neutral mutations.

Analogy with Marbles in a Jar

The process of genetic drift can be illustrated using 20 marbles in a jar to represent 20 organisms in a population. Consider this jar of marbles as the starting population. Half of the marbles in the jar are red and half blue, and both colours correspond to two different alleles of one gene in the population. In each new generation the organisms reproduce at random. To represent this reproduction, randomly select a marble from the original jar and deposit a new marble with the same colour as its "offspring" into a new jar. (The selected marble remains in the original jar.) Repeat this process until there are 20 new marbles in the second jar. The second jar then contains a second generation of "offspring," consisting of 20 marbles of various colours. Unless the second jar contains exactly 10 red marbles and 10 blue marbles, a random shift occurred in the allele frequencies.

Repeat this process a number of times, randomly reproducing each generation of marbles to form the next. The numbers of red and blue marbles picked each generation fluctuates; sometimes more red and sometimes more blue. This fluctuation is analogous to genetic drift – a change in the population's allele frequency resulting from a random variation in the distribution of alleles from one generation to the next.

It is even possible that in any one generation no marbles of a particular colour are chosen, meaning they have no offspring. In this example, if no red marbles are selected, the jar representing the new generation contains only blue offspring. If this happens, the red allele has been lost permanently in the population, while the remaining blue allele has become fixed: all future generations are entirely blue. In small populations, fixation can occur in just a few generations.

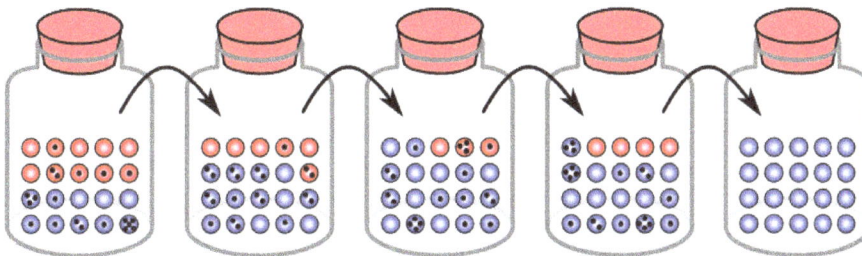

In this simulation, there is fixation in the blue "allele" within five generations.

Probability and allele frequency

The mechanisms of genetic drift can be illustrated with a simplified example. Consider a very large colony of bacteria isolated in a drop of solution. The bacteria are genetically identical except for a single gene with two alleles labeled **A** and **B**. **A** and **B** are neutral alleles meaning that they do not affect the bacteria's ability to survive and reproduce; all bacteria in this colony are equally likely to survive and reproduce. Suppose that half the bacteria have allele **A** and the other half have allele **B**. Thus **A** and **B** each have allele frequency 1/2.

The drop of solution then shrinks until it has only enough food to sustain four bacteria. All other bacteria die without reproducing. Among the four who survive, there are sixteen possible combinations for the **A** and **B** alleles:

(A-A-A-A), (B-A-A-A), (A-B-A-A), (B-B-A-A),

(A-A-B-A), (B-A-B-A), (A-B-B-A), (B-B-B-A),

(A-A-A-B), (B-A-A-B), (A-B-A-B), (B-B-A-B),

(A-A-B-B), (B-A-B-B), (A-B-B-B), (B-B-B-B).

Since all bacteria in the original solution are equally likely to survive when the solution shrinks, the four survivors are a random sample from the original colony. The probability that each of the four survivors has a given allele is 1/2, and so the probability that any particular allele combination occurs when the solution shrinks is

$$\frac{1}{2} \cdot \frac{1}{2} \cdot \frac{1}{2} \cdot \frac{1}{2} = \frac{1}{16}.$$

(The original population size is so large that the sampling effectively happens without replacement). In other words, each of the sixteen possible allele combinations is equally likely to occur, with probability 1/16.

Counting the combinations with the same number of A and B, we get the following table.

A	B	Combinations	Probability
4	0	1	1/16
3	1	4	4/16
2	2	6	6/16
1	3	4	4/16
0	4	1	1/16

As shown in the table, the total number of possible combinations to have an equal (conserved) number of A and B alleles is six, and its probability is 6/16. The total number of possible alternative combinations is ten, and the probability of unequal number of A and B alleles is 10/16. Thus, although the original colony began with an equal number of A and B alleles, chances are that the number of alleles in the remaining population of four members will not be equal. In the latter case, genetic drift has occurred because the population's allele frequencies have changed due to random sampling. In this example the population contracted to just four random survivors, a phenomenon known as population bottleneck.

The probabilities for the number of copies of allele A (or B) that survive (given in the last column of the above table) can be calculated directly from the binomial distribution where the "success" probability (probability of a given allele being present) is 1/2 (i.e., the probability that there are k copies of A (or B) alleles in the combination) is given by

$$\binom{n}{k}\left(\frac{1}{2}\right)^{k}\left(1-\frac{1}{2}\right)^{n-k} = \binom{n}{k}\left(\frac{1}{2}\right)^{n}$$

where $n=4$ is the number of surviving bacteria.

Mathematical Models of Genetic Drift

Mathematical models of genetic drift can be designed using either branching processes or a diffusion equation describing changes in allele frequency in an idealised population.

Wright–Fisher Model

Think of a gene with two alleles, A or B. In diploid populations consisting of N individuals there are 2*N* copies of each gene. An individual can have two copies of the same allele or two different alleles. We can call the frequency of one allele *p* and the frequency of the other *q*. The Wright–Fisher model (named after Sewall Wright and Ronald Fisher) assumes that generations do not overlap (for example, annual plants have exactly one generation per year) and that each copy of the gene found in the new generation is drawn independently at random from all copies of the gene in the old generation. The formula to calculate the probability of obtaining *k* copies of an allele that had frequency *p* in the last generation is then

$$\frac{(2N)!}{k!(2N-k)!} p^k q^{2N-k}$$

where the symbol "!" signifies the factorial function. This expression can also be formulated using the binomial coefficient,

$$\binom{2N}{k} p^k q^{2N-k}$$

Moran Model

The Moran model assumes overlapping generations. At each time step, one individual is chosen to reproduce and one individual is chosen to die. So in each timestep, the number of copies of a given allele can go up by one, go down by one, or can stay the same. This means that the transition matrix is tridiagonal, which means that mathematical solutions are easier for the Moran model than for the Wright–Fisher model. On the other hand, computer simulations are usually easier to perform using the Wright–Fisher model, because fewer time steps need to be calculated. In the Moran model, it takes *N* timesteps to get through one generation, where *N* is the effective population size. In the Wright–Fisher model, it takes just one.

In practice, the Moran model and Wright–Fisher model give qualitatively similar results, but genetic drift runs twice as fast in the Moran model.

Other Models of Drift

If the variance in the number of offspring is much greater than that given by the binomial distribution assumed by the Wright–Fisher model, then given the same overall speed of genetic drift (the variance effective population size), genetic drift is a less powerful force compared to selection. Even for the same variance, if higher moments of the offspring number distribution exceed those of the binomial distribution then again the force of genetic drift is substantially weakened.

Random Effects other than Sampling Error

Random changes in allele frequencies can also be caused by effects other than sampling error, for example random changes in selection pressure.

One important alternative source of stochasticity, perhaps more important than genetic drift, is genetic draft. Genetic draft is the effect on a locus by selection on linked loci. The mathematical properties of genetic draft are different from those of genetic drift. The direction of the random change in allele frequency is autocorrelated across generations.

Drift and Fixation

The Hardy–Weinberg principle states that within sufficiently large populations, the allele frequencies remain constant from one generation to the next unless the equilibrium is disturbed by migration, genetic mutations, or selection.

However, in finite populations, no new alleles are gained from the random sampling of alleles passed to the next generation, but the sampling can cause an existing allele to disappear. Because random sampling can remove, but not replace, an allele, and because random declines or increases in allele frequency influence expected allele distributions for the next generation, genetic drift drives a population towards genetic uniformity over time. When an allele reaches a frequency of 1 (100%) it is said to be "fixed" in the population and when an allele reaches a frequency of 0 (0%) it is lost. Smaller populations achieve fixation faster, whereas in the limit of an infinite population, fixation is not achieved. Once an allele becomes fixed, genetic drift comes to a halt, and the allele frequency cannot change unless a new allele is introduced in the population via mutation or gene flow. Thus even while genetic drift is a random, directionless process, it acts to eliminate genetic variation over time.

Rate of Allele Frequency Change Due to Drift

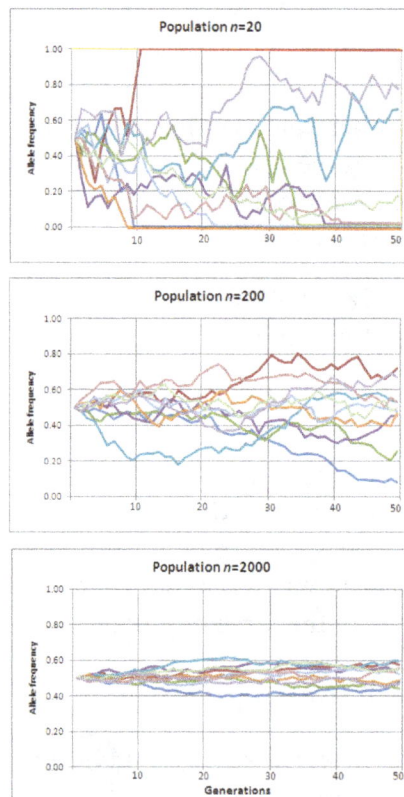

Ten simulations of random genetic drift of a single given allele with an initial frequency distribution 0.5 measured over the course of 50 generations, repeated in three reproductively synchronous populations of different sizes. In these simulations, alleles drift to loss or fixation (frequency of 0.0 or 1.0) only in the smallest population.

Assuming genetic drift is the only evolutionary force acting on an allele, after t generations in many replicated populations, starting with allele frequencies of p and q, the variance in allele frequency across those populations is

$$V_t \approx pq \left(1 - \exp\left(-\frac{t}{2N_e} \right) \right)$$

Time to Fixation or Loss

Assuming genetic drift is the only evolutionary force acting on an allele, at any given time the probability that an allele will eventually become fixed in the population is simply its frequency in the population at that time. For example, if the frequency p for allele **A** is 75% and the frequency q for allele B is 25%, then given unlimited time the probability A will ultimately become fixed in the population is 75% and the probability that B will become fixed is 25%.

The expected number of generations for fixation to occur is proportional to the population size, such that fixation is predicted to occur much more rapidly in smaller populations. Normally the effective population size, which is smaller than the total population, is used to determine these probabilities. The effective population (N_e) takes into account factors such as the level of inbreeding, the stage of the lifecycle in which the population is the smallest, and the fact that some neutral genes are genetically linked to others that are under selection. The effective population size may not be the same for every gene in the same population.

One forward-looking formula used for approximating the expected time before a neutral allele becomes fixed through genetic drift, according to the Wright–Fisher model, is

$$\overline{T}_{\text{fixed}} = \frac{-4N_e(1-p)\ln(1-p)}{p}$$

where T is the number of generations, N_e is the effective population size, and p is the initial frequency for the given allele. The result is the number of generations expected to pass before fixation occurs for a given allele in a population with given size (N_e) and allele frequency (p).

The expected time for the neutral allele to be lost through genetic drift can be calculated as

$$\overline{T}_{\text{lost}} = \frac{-4N_e p}{1-p} \ln p.$$

When a mutation appears only once in a population large enough for the initial frequency to be negligible, the formulas can be simplified to

$$\overline{T}_{\text{fixed}} = 4N_e$$

for average number of generations expected before fixation of a neutral mutation, and

$$\overline{T}_{\text{lost}} = 2 \left(\frac{N_e}{N} \right) \ln(2N)$$

for the average number of generations expected before the loss of a neutral mutation.

Time to loss with both drift and mutation

The formulae above apply to an allele that is already present in a population, and which is subject to neither mutation nor natural selection. If an allele is lost by mutation much more often than it is gained by mutation, then mutation, as well as drift, may influence the time to loss. If the allele prone to mutational loss begins as fixed in the population, and is lost by mutation at rate m per replication, then the expected time in generations until its loss in a haploid population is given by

$$\overline{T}_{\text{lost}} \approx \begin{cases} \dfrac{1}{m}, & \text{if } mN_e \ll 1 \\[2ex] \dfrac{\ln(mN_e)+\gamma}{m} & \text{if } mN_e \gg 1 \end{cases}$$

where γ is equal to Euler's constant. The first approximation represents the waiting time until the first mutant destined for loss, with loss then occurring relatively rapidly by genetic drift, taking time $N_e \ll 1/m$. The second approximation represents the time needed for deterministic loss by mutation accumulation. In both cases, the time to fixation is dominated by mutation via the term $1/m$, and is less affected by the effective population size.

Genetic Drift Versus Natural Selection

In natural populations, genetic drift and natural selection do not act in isolation; both forces are always at play, together with mutation and migration. Neutral evolution is the product of both mutation and drift, not of drift alone. Similarly, even when selection overwhelms genetic drift, it can only act on variation that mutation provides.

While natural selection has a direction, guiding evolution towards heritable adaptations to the current environment, genetic drift has no direction and is guided only by the mathematics of chance. As a result, drift acts upon the genotypic frequencies within a population without regard to their phenotypic effects. In contrast, selection favors the spread of alleles whose phenotypic effects increase survival and/or reproduction of their carriers, lowers the frequencies of alleles that cause unfavorable traits, and ignores those that are neutral.

The law of large numbers predicts that when the absolute number of copies of the allele is small (e.g., in small populations), the magnitude of drift on allele frequencies per generation is larger. The magnitude of drift is large enough to overwhelm selection at any allele frequency when the selection coefficient is less than 1 divided by the effective population size. Non-adaptive evolution resulting from the product of mutation and genetic drift is therefore considered to be a consequential mechanism of evolutionary change primarily within small, isolated populations. The mathematics of genetic drift depend on the effective population size, but it is not clear how this is related to the actual number of individuals in a population. Genetic linkage to other genes that are under selection can reduce the effective population size experienced by a neutral allele. With a higher recombination rate, linkage decreases and with it this local effect on effective population size. This effect is visible in molecular data as a correlation between local recombination rate and genetic diversity, and negative correlation between gene density and diversity at noncoding DNA regions. Stochasticity associated with linkage to other genes that are under selection is not the same as sampling error, and is sometimes known as genetic draft in order to distinguish it from genetic drift.

When the allele frequency is very small, drift can also overpower selection even in large populations. For example, while disadvantageous mutations are usually eliminated quickly in large populations, new advantageous mutations are almost as vulnerable to loss through genetic drift as are neutral mutations. Not until the allele frequency for the advantageous mutation reaches a certain threshold will genetic drift have no effect.

Population Bottleneck

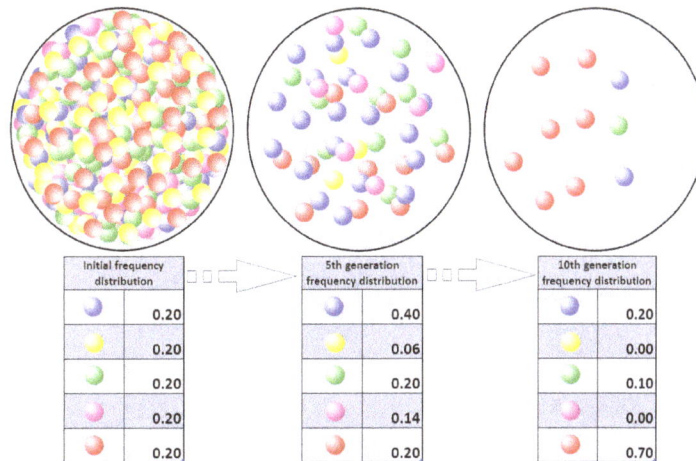

Initial frequency distribution		5th generation frequency distribution		10th generation frequency distribution	
🔵	0.20	🔵	0.40	🔵	0.20
🟡	0.20	🟡	0.06	🟡	0.00
🟢	0.20	🟢	0.20	🟢	0.10
🟣	0.20	🟣	0.14	🟣	0.00
🔴	0.20	🔴	0.20	🔴	0.70

Changes in a population's allele frequency following a population bottleneck: the rapid and radical decline in population size has reduced the population's genetic variation.

A population bottleneck is when a population contracts to a significantly smaller size over a short period of time due to some random environmental event. In a true population bottleneck, the odds for survival of any member of the population are purely random, and are not improved by any particular inherent genetic advantage. The bottleneck can result in radical changes in allele frequencies, completely independent of selection.

The impact of a population bottleneck can be sustained, even when the bottleneck is caused by a one-time event such as a natural catastrophe. An interesting example of a bottleneck causing unusual genetic distribution is the relatively high proportion of individuals with total rod cell color blindness (achromatopsia) on Pingelap atoll in Micronesia. After a bottleneck, inbreeding increases. This increases the damage done by recessive deleterious mutations, in a process known as inbreeding depression. The worst of these mutations are selected against, leading to the loss of other alleles that are genetically linked to them, in a process of background selection. For recessive harmful mutations, this selection can be enhanced as a consequence of the bottleneck, due to genetic purging. This leads to a further loss of genetic diversity. In addition, a sustained reduction in population size increases the likelihood of further allele fluctuations from drift in generations to come.

A population's genetic variation can be greatly reduced by a bottleneck, and even beneficial adaptations may be permanently eliminated. The loss of variation leaves the surviving population vulnerable to any new selection pressures such as disease, climate change or shift in the available food source, because adapting in response to environmental changes requires sufficient genetic variation in the population for natural selection to take place.

There have been many known cases of population bottleneck in the recent past. Prior to the arrival of Europeans, North American prairies were habitat for millions of greater prairie chickens. In Illinois alone, their numbers plummeted from about 100 million birds in 1900 to about 50 birds in the 1990s. The declines in population resulted from hunting and habitat destruction, but the random consequence has been a loss of most of the species' genetic diversity. DNA analysis comparing birds from the mid century to birds in the 1990s documents a steep decline in the genetic variation in just in the latter few decades. Currently the greater prairie chicken is experiencing low reproductive success. However, bottleneck and genetic drift can lead to a genetic loss that increases fitness as seen in Ehrlichia.

Over-hunting also caused a severe population bottleneck in the northern elephant seal in the 19th century. Their resulting decline in genetic variation can be deduced by comparing it to that of the southern elephant seal, which were not so aggressively hunted.

Founder Effect

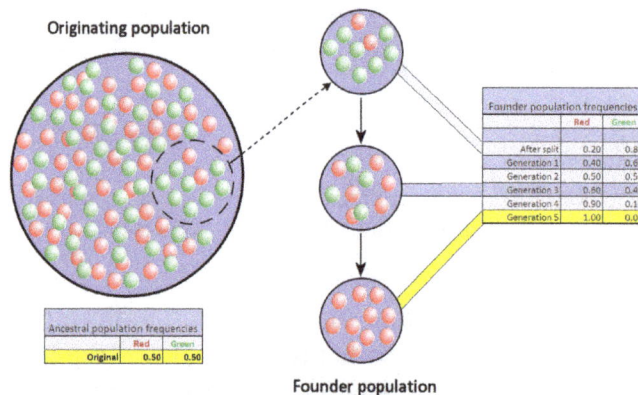

Originating population

Founder population frequencies

	Red	Green
After split	0.20	0.80
Generation 1	0.40	0.60
Generation 2	0.50	0.50
Generation 3	0.60	0.40
Generation 4	0.90	0.10
Generation 5	1.00	0.00

Ancestral population frequencies

	Red	Green
Original	0.50	0.50

Founder population

When very few members of a population migrate to form a separate new population, the founder effect occurs. For a period after the foundation, the small population experiences intensive drift. In the figure this results in fixation of the red allele.

The founder effect is a special case of a population bottleneck, occurring when a small group in a population splinters off from the original population and forms a new one. The random sample of alleles in the just formed new colony is expected to grossly misrepresent the original population in at least some respects. It is even possible that the number of alleles for some genes in the original population is larger than the number of gene copies in the founders, making complete representation impossible. When a newly formed colony is small, its founders can strongly affect the population's genetic make-up far into the future.

A well-documented example is found in the Amish migration to Pennsylvania in 1744. Two members of the new colony shared the recessive allele for Ellis–van Creveld syndrome. Members of the colony and their descendants tend to be religious isolates and remain relatively insular. As a result of many generations of inbreeding, Ellis-van Creveld syndrome is now much more prevalent among the Amish than in the general population.

The difference in gene frequencies between the original population and colony may also trigger the two groups to diverge significantly over the course of many generations. As the difference, or genetic distance, increases, the two separated populations may become distinct, both genetically and phenetically, although not only genetic drift but also natural selection, gene flow, and mutation contribute to this divergence. This potential for relatively rapid changes in the colony's gene frequency led most scientists to consider the founder effect (and by extension, genetic drift) a significant driving force in the evolution of new species. Sewall Wright was the first to attach this significance to random drift and small, newly isolated populations with his shifting balance theory of speciation. Following after Wright, Ernst Mayr created many persuasive models to show that the decline in genetic variation and small population size following the founder effect were critically important for new species to develop. However, there is much less support for this view today since the hypothesis has been tested repeatedly through experimental research and the results have been equivocal at best.

Founder effect was first well-investigated in the USSR by Soviet scientists Lisovskiy V.V., Kuznetsov M.A. and Nikolay Dubinin.

History of the Concept

The concept of genetic drift was first introduced by one of the founders in the field of population genetics, Sewall Wright. His first use of the term "drift" was in 1929, though at the time he was using it in the sense of a directed process of change, or natural selection. Random drift by means of sampling error came to be known as the "Sewall–Wright effect," though he was never entirely comfortable to see his name given to it. Wright referred to all changes in allele frequency as either "steady drift" (e.g., selection) or "random drift" (e.g., sampling error). "Drift" came to be adopted as a technical term in the stochastic sense exclusively. Today it is usually defined still more narrowly, in terms of sampling error, although this narrow definition is not universal. Wright wrote that the "restriction of "random drift" or even "drift" to only one component, the effects of accidents of sampling, tends to lead to confusion." Sewall Wright considered the process of random genetic drift by means of sampling error equivalent to that by means of inbreeding, but later work has shown them to be distinct.

In the early days of the modern evolutionary synthesis, scientists were just beginning to blend the new science of population genetics with Charles Darwin's theory of natural selection. Working within this new framework, Wright focused on the effects of inbreeding on small relatively isolated populations. He introduced the concept of an adaptive landscape in which phenomena such as cross breeding and genetic drift in small populations could push them away from adaptive peaks, which in turn allow natural selection to push them towards new adaptive peaks. Wright thought smaller populations were more suited for natural selection because "inbreeding was sufficiently intense to create new interaction systems through random drift but not intense enough to cause random nonadaptive fixation of genes."

Wright's views on the role of genetic drift in the evolutionary scheme were controversial almost from the very beginning. One of the most vociferous and influential critics was colleague Ronald Fisher. Fisher conceded genetic drift played some role in evolution, but an insignificant one. Fisher has been accused of misunderstanding Wright's views because in his criticisms Fisher seemed to argue Wright had rejected selection almost entirely. To Fisher, viewing the process of evolution as a long, steady, adaptive progression was the only way to explain the ever-increasing complexity from simpler forms. But the debates have continued between the "gradualists" and those who lean more toward the Wright model of evolution where selection and drift together play an important role.

In 1968, Motoo Kimura rekindled the debate with his neutral theory of molecular evolution, which claims that most of the genetic changes are caused by genetic drift acting on neutral mutations.

The role of genetic drift by means of sampling error in evolution has been criticized by John H. Gillespie and William B. Provine, who argue that selection on linked sites is a more important stochastic force.

References

- Kimura, M. (1983). The Neutral Theory of Molecular Evolution. Cambridge University Press, Cambridge. ISBN 0-521-23109-4.

- Robinson, Richard, ed. (2003). "Population Bottleneck". Genetics. 3. New York: Macmillan Reference USA. ISBN 0-02-865609-1. LCCN 2002003560. OCLC 614996575. Retrieved 2015-12-14.

- Staff (2 May 2016). "Researchers find that Earth may be home to 1 trillion species". National Science Foundation. Retrieved 6 May 2016.

- Zimmer, Carl (January 7, 2016). "Genetic Flip Helped Organisms Go From One Cell to Many". The New York Times. Retrieved January 7, 2016.

- Dobzhansky, T; Pavlovsky, O (1957). "An experimental study of interaction between genetic drift and natural selection". Evolution. 11 (3): 311–319. doi:10.2307/2405795. Retrieved 21 April 2016.

- "Using experimental evolution and next-generation sequencing to teach bench and bioinformatic skills". PeerJ PrePrints (3): e1674. 2015. doi:10.7287/peerj.preprints.1356v1.

- Borenstein, Seth (October 19, 2015). "Hints of life on what was thought to be desolate early Earth". Excite. Yonkers, NY: Mindspark Interactive Network. Associated Press. Retrieved 2015-10-20.

- Borenstein, Seth (November 13, 2013). "Oldest fossil found: Meet your microbial mom". Excite. Yonkers, NY: Mindspark Interactive Network. Associated Press. Retrieved 2015-05-31.

- Dougherty, Michael J. (July 20, 1998). "Is the human race evolving or devolving?". Scientific American. Stuttgart: Georg von Holtzbrinck Publishing Group. ISSN 0036-8733. Retrieved 2015-09-11.

- Hyman, Paul (2014). "Bacteriophage as instructional organisms in introductory biology labs". Bacteriophage. 4 (2): e27336. doi:10.4161/bact.27336. ISSN 2159-7081.

- "A Novel Laboratory Activity for Teaching about the Evolution of Multicellularity". The American Biology Teacher. 76 (2): 81–87. 2014. doi:10.1525/abt.2014.76.2.3. ISSN 0002-7685.

- Pearlman, Jonathan (November 13, 2013). "'Oldest signs of life on Earth found'". The Daily Telegraph. London: Telegraph Media Group. Retrieved 2014-12-15.

- Novacek, Michael J. (November 8, 2014). "Prehistory's Brilliant Future". The New York Times. New York: The New York Times Company. ISSN 0362-4331. Retrieved 2014-12-25.

- Scott-Phillips, T. C., Laland, K. N., Shuker, D. M., Dickins, T. E. and West, S. A. (2014). "The Niche Construction Perspective: A Critical Appraisal". Evolution 68: 1231-1243.

Evolution of Biology

Biology concerns itself with the study of life, evolution and distribution of a living organism.The etymological meaning of biology in Greek is life; the origins of modern biology can easily be traced back to the Greeks. The chapter serves as a source to understand the history on the evolution of biology, although modern biology is a relatively recent development, sciences included within it have been studied since ancient times.

History of Evolutionary Thought

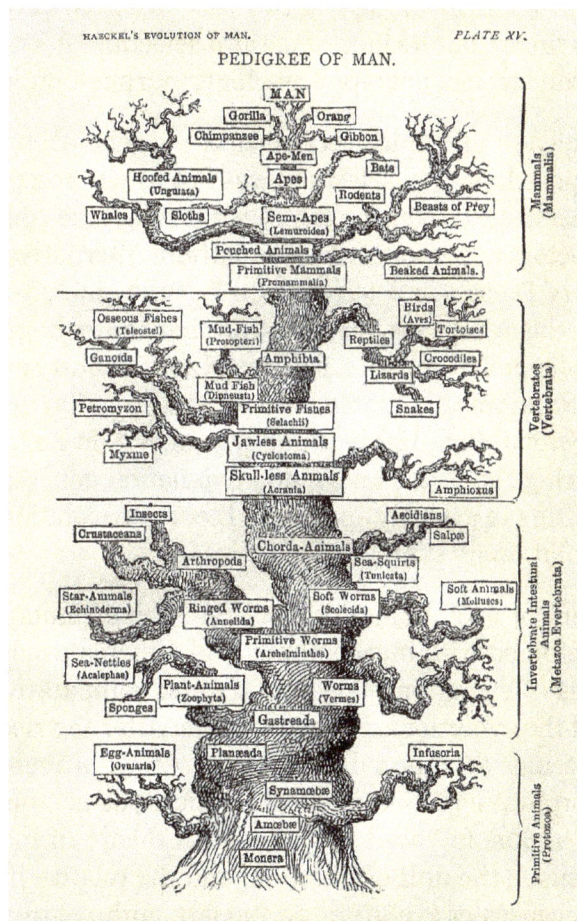

The tree of life as depicted by Ernst Haeckel in The Evolution of Man (1879) illustrates the 19th-century view of evolution as a progressive process leading towards man.

Evolutionary thought, the conception that species change over time, has roots in antiquity - in the ideas of the ancient Greeks, Romans, and Chinese as well as in medieval Islamic science. With the beginnings of modern biological taxonomy in the late 17th century, two opposed ideas influenced Western biological thinking:

- essentialism, the belief that every species has essential characteristics that are unalterable, a concept which had developed from medieval Aristotelian metaphysics, and that fit well with natural theology

- the development of the new anti-Aristotelian approach to modern science: as the Enlightenment progressed, evolutionary cosmology and the mechanical philosophy spread from the physical sciences to natural history

Naturalists began to focus on the variability of species; the emergence of paleontology with the concept of extinction further undermined static views of nature. In the early 19th century Jean-Baptiste Lamarck (1744 – 1829) proposed his theory of the transmutation of species, the first fully formed theory of evolution.

In 1858 Charles Darwin and Alfred Russel Wallace published a new evolutionary theory, explained in detail in Darwin's *On the Origin of Species* (1859). Unlike Lamarck, Darwin proposed common descent and a branching tree of life, meaning that two very different species could share a common ancestor. Darwin based his theory on the idea of natural selection: it synthesized a broad range of evidence from animal husbandry, biogeography, geology, morphology, and embryology.

Debate over Darwin's work led to the rapid acceptance of the general concept of evolution, but the specific mechanism he proposed, natural selection, was not widely accepted until it was revived by developments in biology that occurred during the 1920s through the 1940s. Before that time most biologists regarded other factors as responsible for evolution. Alternatives to natural selection suggested during "the eclipse of Darwinism" (circa 1880 to 1920) included inheritance of acquired characteristics (neo-Lamarckism), an innate drive for change (orthogenesis), and sudden large mutations (saltationism). Mendelian genetics, a series of 19th Century experiments with pea plant variations rediscovered in 1900, was integrated with natural selection by Ronald Fisher during the 1910s to 1930s, and along with J. B. S. Haldane and Sewall Wright he founded the new discipline of population genetics. During the 1930s and 1940s population genetics became integrated with other biological fields, resulting in a widely applicable theory of evolution that encompassed much of biology—the modern evolutionary synthesis.

Following the establishment of evolutionary biology, studies of mutation and genetic diversity in natural populations, combined with biogeography and systematics, led to sophisticated mathematical and causal models of evolution. Paleontology and comparative anatomy allowed more detailed reconstructions of the evolutionary history of life. After the rise of molecular genetics in the 1950s, the field of molecular evolution developed, based on protein sequences and immunological tests, and later incorporating RNA and DNA studies. The gene-centered view of evolution rose to prominence in the 1960s, followed by the neutral theory of molecular evolution, sparking debates over adaptationism, the unit of selection, and the relative importance of genetic drift versus natural selection as causes of evolution. In the late 20th-century, DNA sequencing led to molecular phylogenetics and the reorganization of the tree of life into the three-domain system by Carl Woese. In addition, the newly recognized factors of symbiogenesis and horizontal gene transfer introduced yet more complexity into evolutionary theory. Discoveries in evolutionary biology have made a significant impact not just within the traditional branches of biology, but also in other academic disciplines (for example: anthropology and psychology) and on society at large.

Antiquity

Greeks

Proposals that one type of animal, even humans, could descend from other types of animals, are known to go back to the first pre-Socratic Greek philosophers. Anaximander of Miletus (c. 610 – 546 BC) proposed that the first animals lived in water, during a wet phase of the Earth's past, and that the first land-dwelling ancestors of mankind must have been born in water, and only spent part of their life on land. He also argued that the first human of the form known today must have been the child of a different type of animal, because man needs prolonged nursing to live. Empedocles (c. 490 – 430 BC), argued that what we call birth and death in animals are just the mingling and separations of elements which cause the countless "tribes of mortal things." Specifically, the first animals and plants were like disjointed parts of the ones we see today, some of which survived by joining in different combinations, and then intermixing, and wherever "everything turned out as it would have if it were on purpose, there the creatures survived, being accidentally compounded in a suitable way." Other philosophers who became more influential in the Middle Ages, including Plato (c. 428/427 – 348/347 BC), Aristotle (384 – 322 BC), and members of the Stoic school of philosophy, believed that the species of all things, not only living things, were fixed by divine design.

Plato (left) and Aristotle (right), a detail from *The School of Athens* (1509–1511) by Raphael

Plato was called by biologist Ernst Mayr "the great antihero of evolutionism," because he promoted belief in essentialism, which is also referred to as the theory of Forms. This theory holds that each natural type of object in the observed world is an imperfect manifestation of the ideal, form or "species" which defines that type. In his *Timaeus* for example, Plato has a character tell a story that the Demiurge created the cosmos and everything in it because, being good, and hence, "... free from jealousy, He desired that all things should be as like Himself as they could be." The creator created all conceivable forms of life, since "... without them the universe will be incomplete, for it will not contain every kind of animal which it ought to contain, if it is to be perfect." This "principle of plenitude"—the idea that all potential forms of life are essential to a perfect creation—greatly influenced Christian thought. However some historians of science have questioned how much influence Plato's essentialism had on natural philosophy by stating that many philosophers after Plato believed that species might be capable of transformation and that the idea that biologic species were fixed and possessed unchangeable essential characteristics did not become important until the beginning of biological taxonomy in the 17th and 18th centuries.

Aristotle, the most influential of the Greek philosophers in Europe in the Middle Ages, was a student of Plato and is also the earliest natural historian whose work has been preserved in any real detail. His writings on biology resulted from his research into natural history on and around the island of Lesbos, and have survived in the form of four books, usually known by their Latin names, *De anima* (*On the Soul*), *Historia animalium* (*History of Animals*), *De generatione animalium* (*Generation of Animals*), and *De partibus animalium* (*On the Parts of Animals*). Aristotle's works contain some remarkably astute observations and interpretations—along with sundry myths and mistakes—reflecting the uneven state of knowledge during his time. However, for Charles Singer, "Nothing is more remarkable than [Aristotle's] efforts to [exhibit] the relationships of living things as a *scala naturae*." This *scala naturae*, described in *Historia animalium*, classified organisms in relation to a hierarchical "Ladder of Life" or "great chain of being," placing them according to their complexity of structure and function, with organisms that showed greater vitality and ability to move described as "higher organisms." Aristotle believed that features of living organisms showed clearly that they must have had what he called a final cause, that is to say that they had been designed for a purpose. He explicitly rejected the view of Empedocles that living creatures might have originated by chance.

Other Greek philosophers, such as Zeno of Citium (334 – 262 BC) the founder of the Stoic school of philosophy, agreed with Aristotle and other earlier philosophers that nature showed clear evidence of being designed for a purpose; this view is known as teleology. The Roman Stoic philosopher Cicero (106 – 43 BC) wrote that Zeno was known to have held the view, central to Stoic physics, that nature is primarily "directed and concentrated...to secure for the world...the structure best fitted for survival."

Epicurus (341 – 270 BC) anticipated the idea of natural selection. The Roman philosopher and atomist Lucretius (c. 99 – 55 BC) explicated these ideas in his poem *De rerum natura* (*On the Nature of Things*). In the Epicurean system, it was assumed that many species had been spontaneously generated from Gaia in the past, but that only the most functional forms survived to have offspring. The Epicureans do not seem to have anticipated the full theory of evolution as we now know it and seem to have postulated separate abiogenetic events for each species rather than postulating a single abiogenetic event coupled with the differentiation of species over time from a single (or small number of) originating parent organism(s).

Chinese

Ancient Chinese thinkers such as Zhuang Zhou (c. 369 – 286 BC), a Taoist philosopher, expressed ideas on changing biologic species. According to Joseph Needham, Taoism explicitly denies the fixity of biological species and Taoist philosophers speculated that species had developed differing attributes in response to differing environments. Taoism regards humans, nature and the heavens as existing in a state of "constant transformation" known as the *Tao*, in contrast with the more static view of nature typical of Western thought.

Romans

Lucretius' poem *De rerum natura* provides the best surviving explanation of the ideas of the Greek Epicurean philosophers. It describes the development of the cosmos, the Earth, living things, and human society through purely naturalistic mechanisms, without any reference to supernatural involvement. *De rerum natura* would influence the cosmological and evolutionary speculations of philosophers and scientists during and after the Renaissance. This view was in strong contrast with the views of Roman philosophers of the Stoic school such as Cicero, Seneca the Younger (c. 4 BC – AD 65), and Pliny the Elder (23 – 79 AD) who had a strongly teleological view of the natural world that influenced Christian theology. Cicero reports that the peripatetic and Stoic view of nature as an agency concerned most basically with producing life "best fitted for survival" was taken for granted among the Hellenistic elite.

Augustine of Hippo

In line with earlier Greek thought, the 4th-century bishop and theologian, Augustine of Hippo, wrote that the creation story in the Book of Genesis should not be read too literally. In his book *De Genesi ad litteram* (*On the Literal Meaning of Genesis*), he stated that in some cases new creatures may have come about through the "decomposition" of earlier forms of life. For Augustine, "plant, fowl and animal life are not perfect ... but created in a state of potentiality," unlike what he considered the theologically perfect forms of angels, the firmament and the human soul. Augustine's idea 'that forms of life had been transformed "slowly over time"' prompted Father Giuseppe Tanzella-Nitti, Professor of Theology at the Pontifical Santa Croce University in Rome, to claim that Augustine had suggested a form of evolution.

Henry Fairfield Osborn wrote in *From the Greeks to Darwin* (1894):

"If the orthodoxy of Augustine had remained the teaching of the Church, the final establishment of Evolution would have come far earlier than it did, certainly during the eighteenth instead of the nineteenth century, and the bitter controversy over this truth of Nature would never have arisen. ...Plainly as the direct or instantaneous Creation of animals and plants appeared to be taught in Genesis, Augustine read this in the light of primary causation and the gradual development from the imperfect to the perfect of Aristotle. This most influential teacher thus handed down to his followers opinions which closely conform to the progressive views of those theologians of the present day who have accepted the Evolution theory."

In *A History of the Warfare of Science with Theology in Christendom* (1896), Andrew Dickson White wrote about Augustine's attempts to preserve the ancient evolutionary approach to the creation as follows:

"For ages a widely accepted doctrine had been that water, filth, and carrion had received power from the Creator to generate worms, insects, and a multitude of the smaller animals; and this doctrine had been especially welcomed by St. Augustine and many of the fathers, since it relieved the Almighty of making, Adam of naming, and Noah of living in the ark with these innumerable despised species."

In Augustine's *De Genesi contra Manichæos*, on Genesis he says: "To suppose that God formed man from the dust with bodily hands is very childish. ...God neither formed man with bodily hands nor did he breathe upon him with throat and lips." Augustine suggests in other work his theory of the later development of insects out of carrion, and the adoption of the old emanation or evolution theory, showing that "certain very small animals may not have been created on the fifth and sixth days, but may have originated later from putrefying matter." Concerning Augustine's *De Trinitate* (*On the Trinity*), White wrote that Augustine "...develops at length the view that in the creation of living beings there was something like a growth—that God is the ultimate author, but works through secondary causes; and finally argues that certain substances are endowed by God with the power of producing certain classes of plants and animals."

Middle Ages

Islamic Philosophy and the Struggle for Existence

A page from the *Kitāb al-Hayawān* (English: *Book of Animals*) by al-Jāḥiẓ

Although Greek and Roman evolutionary ideas died out in Europe after the fall of the Roman Empire, they were not lost to Islamic philosophers and scientists. In the Islamic Golden Age of the 8th to the 13th centuries, philosophers explored ideas about natural history. These ideas included transmutation from non-living to living: "from mineral to plant, from plant to animal, and from animal to man."

In the medieval Islamic world, the scholar al-Jāḥiẓ (776 – c. 868) wrote his *Book of Animals* in the 9th century. Conway Zirkle, writing about the history of natural selection in 1941, said that an excerpt from this work was the only relevant passage he had found from an Arabian scholar. He provided a quotation describing the struggle for existence, citing a Spanish translation of this work: "The rat goes out for its food, and is clever in getting it, for it eats all animals inferior to it in strength," and in turn, it "has to avoid snakes and birds and serpents of prey, who look for it in order to devour it" and are stronger than the rat. Mosquitoes "know instinctively that blood is the thing which makes them live" and when they see an animal, "they know that the skin has been fashioned to serve them as food." In turn, flies hunt the mosquito "which is the food that they like best," and predators eat the flies. "All animals, in short, can not exist without food, neither can the hunting animal escape being hunted in his turn. Every weak animal devours those weaker than itself. Strong animals cannot escape being devoured by other animals stronger than they. And in this respect, men do not differ from animals, some with respect to others, although they do not arrive at the same extremes. In short, God has disposed some human beings as a cause of life for others, and likewise, he has disposed the latter as a cause of the death of the former." Al-Jāḥiẓ also wrote descriptions of food chains.

Some of Ibn Khaldūn's thoughts, according to some commentators, anticipate the biological theory of evolution. In 1377, Ibn Khaldūn wrote the *Muqaddimah* in which he asserted that humans developed from "the world of the monkeys," in a process by which "species become more numerous" In chapter 1 he writes: "This world with all the created things in it has a certain order and solid construction. It shows nexuses between causes and things caused, combinations of some parts of creation with others, and transformations of some existent things into others, in a pattern that is both remarkable and endless."

The *Muqaddimah* also states in chapter 6:

"We explained there that the whole of existence in (all) its simple and composite worlds is arranged in a natural order of ascent and descent, so that everything constitutes an uninterrupted continuum. The essences at the end of each particular stage of the worlds are by nature prepared to be transformed into the essence adjacent to them, either above or below them. This is the case with the simple material elements; it is the case with palms and vines, (which constitute) the last stage of plants, in their relation to snails and shellfish, (which constitute) the (lowest) stage of animals. It is also the case with monkeys, creatures combining in themselves cleverness and perception, in their relation to man, the being who has the ability to think and to reflect. The preparedness (for transformation) that exists on either side, at each stage of the worlds, is meant when (we speak about) their connection."

Nasīr al-Dīn Tūsī

In his Akhlaq-i-Nasri, Tusi put forward a basic theory for the evolution of species almost 600 years before Charles Darwin, the English naturalist credited with advancing the idea, was born. He begins his theory of evolution with the universe once consisting of equal and similar elements. According to Tusi, internal contradictions began appearing, and as a result, some substances began developing faster and differently from other substances. He then explains how the elements evolved into minerals, then plants, then animals, and then humans. Tusi then goes on to explain how hereditary variability was an important factor for biological evolution of living things:

"The organisms that can gain the new features faster are more variable. As a result, they gain advantages over other creatures. [...] The bodies are changing as a result of the internal and external interactions."

Tusi discusses how organisms are able to adapt to their environments:

"Look at the world of animals and birds. They have all that is necessary for defense, protection and daily life, including strengths, courage and appropriate tools [organs] [...] Some of these organs are real weapons, [...] For example, horns-spear, teeth and claws-knife and needle, feet and hoofs-cudgel. The thorns and needles of some animals are similar to arrows. [...] Animals that have no other means of defense (as the gazelle and fox) protect themselves with the help of flight and cunning. [...] Some of them, for example, bees, ants and some bird species, have united in communities in order to protect themselves and help each other."

Tusi recognized three types of living things: plants, animals, and humans. He wrote:

"Animals are higher than plants, because they are able to move consciously, go after food, find and eat useful things. [...] There are many differences between the animal and plant species, [...] First of all, the animal kingdom is more complicated. Besides, reason is the most beneficial feature of animals. Owing to reason, they can learn new things and adopt new, non-inherent abilities. For example, the trained horse or hunting falcon...is at a higher point of development in the animal world. The first steps of human perfection begin from here."

Tusi then explains how humans evolved from advanced animals:

"Such humans [probably anthropoid apes] live in the Western Sudan and other distant corners of the world. They are close to animals by their habits, deeds and behavior. [...] The human has features that distinguish him from other creatures, but he has other features that unite him with the animal world, vegetable kingdom or even with the inanimate bodies. [...] Before [the creation of humans], all differences between organisms were of the natural origin. The next step will be associated with spiritual perfection, will, observation and knowledge. [...] All these facts prove that the human being is placed on the middle step of the evolutionary stairway. According to his inherent nature, the human is related to the lower beings, and only with the help of his will can he reach the higher development level."

Christian Philosophy and the Great Chain of Being

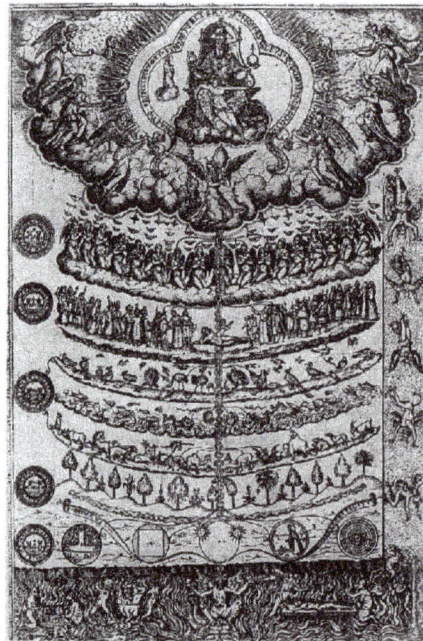

Drawing of the great chain of being from *Rhetorica Christiana* (English: *Christian Rhetoric*) (1579) by Diego Valadés

During the Early Middle Ages, Greek classical learning was all but lost to the West. However, contact with the Islamic world, where Greek manuscripts were preserved and expanded, soon led to a massive spate of Latin translations in the 12th century. Europeans were re-introduced to the works of Plato and Aristotle, as well as to Islamic thought. Christian thinkers of the scholastic school, in particular Peter Abelard (1079 – 1142) and Thomas Aquinas (1225 – 1274), combined Aristotelian classification with Plato's ideas of the goodness of God, and of all potential life forms being present in a perfect creation, to organize all inanimate, animate, and spiritual beings into a huge interconnected system: the *scala naturae*, or great chain of being.

Within this system, everything that existed could be placed in order, from "lowest" to "highest," with Hell at the bottom and God at the top—below God, an angelic hierarchy marked by the orbits of the planets, mankind in an intermediate position, and worms the lowest of the animals. As the universe was ultimately perfect, the great chain of being was also perfect. There were no empty links in the chain, and no link was represented by more than one species. Therefore, no species could ever move from one position to another. Thus, in this Christianized version of Plato's perfect universe, species could never change, but remained forever fixed, in accordance with the text of the Book of Genesis. For humans to forget their position was seen as sinful, whether they behaved like lower animals or aspired to a higher station than was given them by their Creator.

Creatures on adjacent steps were expected to closely resemble each other, an idea expressed in the saying: *natura non facit saltum* ("nature does not make leaps"). This basic concept of the great chain of being greatly influenced the thinking of Western civilization for centuries (and still has an influence today). It formed a part of the argument from design presented by natural theology. As a classification system, it became the major organizing principle and foundation of the emerging science of biology in the 17th and 18th centuries.

Thomas Aquinas on Creation and Natural Processes

While the development of the great chain of being and the argument from design by Christian theologians contributed to the view that the natural world fit into an unchanging designed hierarchy, some theologians were more open to the possibility that the world might have developed through natural processes. Thomas Aquinas went even farther than Augustine of Hippo in arguing that scriptural texts like Genesis should not be interpreted in a literal way that conflicted with or constrained what natural philosophers learned about the workings of the natural world. He felt that the autonomy of nature was a sign of God's goodness and that there was no conflict between the concept of a divinely created universe, and the idea that the universe may have evolved over time through natural mechanisms. However, Aquinas disputed the views of those like the ancient Greek philosopher Empedocles who held that such natural processes showed that the universe could have developed without an underlying purpose. Rather holding that: "Hence, it is clear that nature is nothing but a certain kind of art, i.e., the divine art, impressed upon things, by which these things are moved to a determinate end. It is as if the shipbuilder were able to give to timbers that by which they would move themselves to take the form of a ship."

Renaissance and Enlightenment

In the first half of the 17th century, René Descartes' mechanical philosophy encouraged the use of the metaphor of the universe as a machine, a concept that would come to characterise the scientific revolution. Between 1650 and 1800, some naturalists, such as Benoît de Maillet, produced theories that maintained that the universe, the Earth, and life, had developed mechanically, without divine guidance. In contrast, most contemporary theories of evolution, such of those of Gottfried Leibniz and Johann Gottfried Herder, regarded evolution as a fundamentally *spiritual* process. In 1751, Pierre Louis Maupertuis veered toward more materialist ground. He wrote of natural modifications occurring during reproduction and accumulating over the course of many generations, producing races and even new species, a description that anticipated in general terms the concept of natural selection.

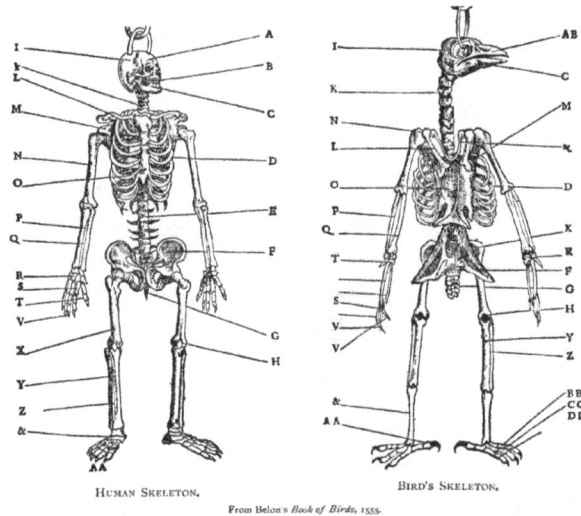

Pierre Belon compared the skeletons of humans (left) and birds (right) in his *L'Histoire de la nature des oyseaux* (English: *The Natural History of Birds*) (1555)

Maupertuis' ideas were in opposition to the influence of early taxonomists like John Ray. In the late 17th century, Ray had given the first formal definition of a biological species, which he described as being characterized by essential unchanging features, and stated the seed of one species could never give rise to another. The ideas of Ray and other 17th-century taxonomists were influenced by natural theology and the argument from design.

The word *evolution* (from the Latin *evolutio*, meaning "to unroll like a scroll") was initially used to refer to embryological development; its first use in relation to development of species came in 1762, when Charles Bonnet used it for his concept of "pre-formation," in which females carried a miniature form of all future generations. The term gradually gained a more general meaning of growth or progressive development.

Later in the 18th century, the French philosopher Georges-Louis Leclerc, Comte de Buffon, one of the leading naturalists of the time, suggested that what most people referred to as species were really just well-marked varieties, modified from an original form by environmental factors. For example, he believed that lions, tigers, leopards and house cats might all have a common ancestor. He further speculated that the 200 or so species of mammals then known might have descended from as few as 38 original animal forms. Buffon's evolutionary ideas were limited; he believed each of the original forms had arisen through spontaneous generation and that each was shaped by "internal moulds" that limited the amount of change. Buffon's works, *Histoire naturelle* (1749–1789) and *Époques de la nature* (1778), containing well-developed theories about a completely materialistic origin for the Earth and his ideas questioning the fixity of species, were extremely influential. Another French philosopher, Denis Diderot, also wrote that living things might have first arisen through spontaneous generation, and that species were always changing through a constant process of experiment where new forms arose and survived or not based on trial and error; an idea that can be considered a partial anticipation of natural selection. Between 1767 and 1792, James Burnett, Lord Monboddo, included in his writings not only the concept that man had descended from primates, but also that, in response to the environment, creatures had found methods of transforming their characteristics over long time intervals. Charles Darwin's grandfather, Erasmus Darwin, published *Zoonomia* (1794–1796) which suggested that "all warm-blooded animals have arisen from one living filament." In his poem *Temple of Nature* (1803), he described the rise of life from minute organisms living in mud to all of its modern diversity.

Early 19th Century

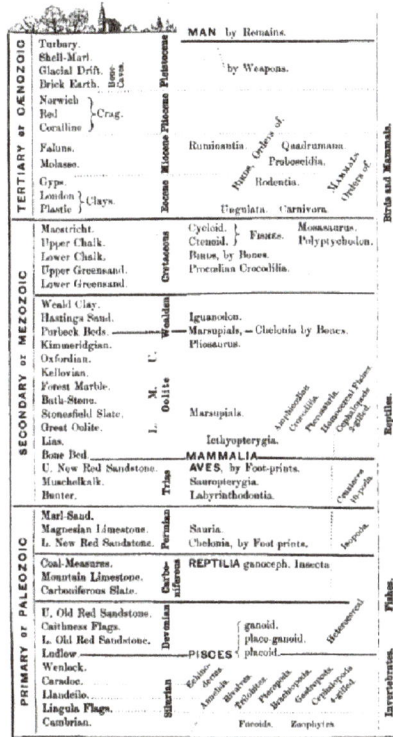

Diagram of the geologic timescale from *Palæontology* (1861) by Richard Owen showing the appearance of major animal types

Paleontology and Geology

In 1796, Georges Cuvier published his findings on the differences between living elephants and those found in the fossil record. His analysis identified mammoths and mastodons as distinct species, different from any living animal, and effectively ended a long-running debate over whether a species could become extinct. In 1788, James Hutton described gradual geological processes operating continuously over deep time. In the 1790s, William Smith began the process of ordering rock strata by examining fossils in the layers while he worked on his geologic map of England. Independently, in 1811, Cuvier and Alexandre Brongniart published an influential study of the geologic history of the region around Paris, based on the stratigraphic succession of rock layers. These works helped establish the antiquity of the Earth. Cuvier advocated catastrophism to explain the patterns of extinction and faunal succession revealed by the fossil record.

Knowledge of the fossil record continued to advance rapidly during the first few decades of the 19th century. By the 1840s, the outlines of the geologic timescale were becoming clear, and in 1841 John Phillips named three major eras, based on the predominant fauna of each: the Paleozoic, dominated by marine invertebrates and fish, the Mesozoic, the age of reptiles, and the current Cenozoic age of mammals. This progressive picture of the history of life was accepted even by conservative English geologists like Adam Sedgwick and William Buckland; however, like Cuvier, they attributed the progression to repeated catastrophic episodes of extinction followed by new episodes of creation. Unlike Cuvier, Buckland and some other advocates of natural theology among British geologists made efforts to explicitly link the last catastrophic episode proposed by Cuvier to the biblical flood.

From 1830 to 1833, geologist Charles Lyell published his multi-volume work *Principles of Geology*, which, building on Hutton's ideas, advocated a uniformitarian alternative to the catastrophic theory of geology. Lyell claimed that, rather than being the products of cataclysmic (and possibly supernatural) events, the geologic features of the Earth are better explained as the result of the same gradual geologic forces observable in the present day—but acting over immensely long periods of time. Although Lyell opposed evolutionary ideas (even questioning the consensus that the fossil record demonstrates a true progression), his concept that the Earth was shaped by forces working gradually over an extended period, and the immense age of the Earth assumed by his theories, would strongly influence future evolutionary thinkers such as Charles Darwin.

Transmutation of Species

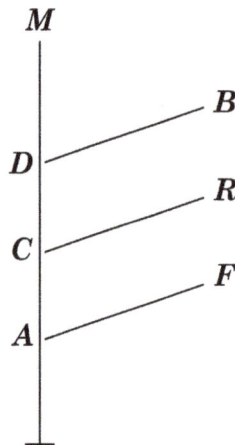

Diagram from *Vestiges of the Natural History of Creation* (1844) by Robert Chambers shows a model of development where fish (F), reptiles (R), and birds (B) represent branches from a path leading to mammals (M)

Jean-Baptiste Lamarck proposed, in his *Philosophie Zoologique* of 1809, a theory of the transmutation of species ("transformisme"). Lamarck did not believe that all living things shared a common ancestor but rather that simple forms of life were created continuously by spontaneous generation. He also believed that an innate life force drove species to become more complex over time, advancing up a linear ladder of complexity that was related to the great chain of being. Lamarck recognized that species adapted to their environment. He explained this by saying that the same innate force driving increasing complexity caused the organs of an animal (or a plant) to change based on the use or disuse of those organs, just as exercise affects muscles. He argued that these changes would be inherited by the next generation and produce slow adaptation to the environment. It was this secondary mechanism of adaptation through the inheritance of acquired characteristics that would become known as Lamarckism and would influence discussions of evolution into the 20th century.

A radical British school of comparative anatomy that included the anatomist Robert Edmond Grant was closely in touch with Lamarck's French school of *Transformationism*. One of the French scientists who influenced Grant was the anatomist Étienne Geoffroy Saint-Hilaire, whose ideas on the unity of various animal body plans and the homology of certain anatomical structures would be widely influential and lead to intense debate with his colleague Georges Cuvier. Grant became an authority on the anatomy and reproduction of marine invertebrates. He developed Lamarck's and Erasmus Darwin's ideas of transmutation and evolutionism, and investigated homology, even proposing that plants and animals had a common evolutionary starting point. As a young student, Charles Darwin joined Grant in investigations of the life cycle of marine animals. In 1826, an anonymous paper, probably written by Robert Jameson, praised Lamarck for explaining how higher animals had "evolved" from the simplest worms; this was the first use of the word "evolved" in a modern sense.

In 1844, the Scottish publisher Robert Chambers anonymously published an extremely controversial but widely read book entitled *Vestiges of the Natural History of Creation*. This book proposed an evolutionary scenario for the origins of the Solar System and of life on Earth. It claimed that the fossil record showed a progressive ascent of animals, with current animals branching off a main line that leads progressively to humanity. It implied that the transmutations lead to the unfolding of a preordained plan that had been woven into the laws that governed the universe. In this sense it was less completely materialistic than the ideas of radicals like Grant, but its implication that humans were only the last step in the ascent of animal life incensed many conservative thinkers. The high profile of the public debate over *Vestiges*, with its depiction of evolution as a progressive process, would greatly influence the perception of Darwin's theory a decade later.

Ideas about the transmutation of species were associated with the radical materialism of the Enlightenment and were attacked by more conservative thinkers. Cuvier attacked the ideas of Lamarck and Geoffroy, agreeing with Aristotle that species were immutable. Cuvier believed that the individual parts of an animal were too closely correlated with one another to allow for one part of the anatomy to change in isolation from the others, and argued that the fossil record showed patterns of catastrophic extinctions followed by repopulation, rather than gradual change over time. He also noted that drawings of animals and animal mummies from Egypt, which were thousands of years old, showed no signs of change when compared with modern animals. The strength of Cuvier's arguments and his scientific reputation helped keep transmutational ideas out of the mainstream for decades.

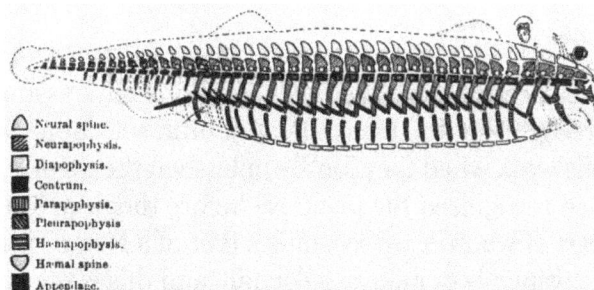

△ Neural spine.
▨ Neurapophysis.
▢ Diapophysis.
■ Centrum.
▥ Parapophysis.
▨ Pleurapophysis.
▤ Hæmapophysis.
▽ Hæmal spine.
■ Appendage.

This 1848 diagram by Richard Owen shows his conceptual archetype for all vertebrates.

In Great Britain, the philosophy of natural theology remained influential. William Paley's 1802 book *Natural Theology* with its famous watchmaker analogy had been written at least in part as a response to the transmutational ideas of Erasmus Darwin. Geologists influenced by natural theology, such as Buckland and Sedgwick, made a regular practice of attacking the evolutionary ideas of Lamarck, Grant, and *Vestiges*. Although Charles Lyell opposed scriptural geology, he also believed in the immutability of species, and in his *Principles of Geology*, he criticized Lamarck's theories of development. Idealists such as Louis Agassiz and Richard Owen believed that each species was fixed and unchangeable because it represented an idea in the mind of the creator. They believed that relationships between species could be discerned from developmental patterns in embryology, as well as in the fossil record, but that these relationships represented an underlying pattern of divine thought, with progressive creation leading to increasing complexity and culminating in humanity. Owen developed the idea of "archetypes" in the Divine mind that would produce a sequence of species related by anatomical homologies, such as vertebrate limbs. Owen led a public campaign that successfully marginalized Grant in the scientific community. Darwin would make good use of the homologies analyzed by Owen in his own theory, but the harsh treatment of Grant, and the controversy surrounding *Vestiges*, showed him the need to ensure that his own ideas were scientifically sound.

Anticipations of Natural Selection

It is possible to look through the history of biology from the ancient Greeks onwards and discover anticipations of almost all of Charles Darwin's key ideas. For example, Loren Eiseley has found isolated passages written by Buffon suggesting he was almost ready to piece together a theory of natural selection, but such anticipations should not be taken out of the full context of the writings or of cultural values of the time which could make Darwinian ideas of evolution unthinkable.

When Darwin was developing his theory, he investigated selective breeding and was impressed by Sebright's observation that "A severe winter, or a scarcity of food, by destroying the weak and the unhealthy, has all the good effects of the most skilful selection" so that "the weak and the unhealthy do not live to propagate their infirmities." Darwin was influenced by Charles Lyell's ideas of environmental change causing ecological shifts, leading to what Augustin de Candolle had called a war between competing plant species, competition well described by the botanist William Herbert. Darwin was struck by Thomas Robert Malthus' phrase "struggle for existence" used of warring human tribes.

Several writers anticipated evolutionary aspects of Darwin's theory, and in the third edition of *On the Origin of Species* published in 1861 Darwin named those he knew about in an introductory appendix, *An Historical Sketch of the Recent Progress of Opinion on the Origin of Species*, which he expanded in later editions.

In 1813, William Charles Wells read before the Royal Society essays assuming that there had been evolution of humans, and recognising the principle of natural selection. Darwin and Alfred Russel Wallace were unaware of this work when they jointly published the theory in 1858, but Darwin later acknowledged that Wells had recognised the principle before them, writing that the paper "An Account of a White Female, part of whose Skin resembles that of a Negro" was published in 1818, and "he distinctly recognises the principle of natural selection, and this is the first recognition which has been indicated; but he applies it only to the races of man, and to certain characters alone."

Patrick Matthew wrote in the obscure book *On Naval Timber and Arboriculture* (1831) of "continual balancing of life to circumstance. ... [The] progeny of the same parents, under great differences of circumstance, might, in several generations, even become distinct species, incapable of co-reproduction." Charles Darwin discovered this work after the initial publication of the *Origin*. In the brief historical sketch that Darwin included in the 3rd edition he says "Unfortunately the view was given by Mr. Matthew very briefly in scattered passages in an Appendix to a work on a different subject ... He clearly saw, however, the full force of the principle of natural selection."

However, as historian of science Peter J. Bowler says, "Through a combination of bold theorizing and comprehensive evaluation, Darwin came up with a concept of evolution that was unique for the time." Bowler goes on to say that simple priority alone is not enough to secure a place in the history of science; someone has to develop an idea and convince others of its importance to have a real impact. Thomas Henry Huxley said in his essay on the reception of *On the Origin of Species*:

"The suggestion that new species may result from the selective action of external conditions upon the variations from their specific type which individuals present—and which we call "spontaneous," because we are ignorant of their causation—is as wholly unknown to the historian of scientific ideas as it was to biological specialists before 1858. But that suggestion is the central idea of the 'Origin of Species,' and contains the quintessence of Darwinism."

Charles Darwin's first sketch of an evolutionary tree from his "B" notebook on the
transmutation of species (1837–1838)

Natural Selection

The biogeographical patterns Charles Darwin observed in places such as the Galápagos Islands during the second voyage of HMS *Beagle* caused him to doubt the fixity of species, and in 1837 Darwin started the first of a series of secret notebooks on transmutation. Darwin's observations led him to view transmutation as a process of divergence and branching, rather than the ladder-like progression envisioned by Jean-Baptiste Lamarck and others. In 1838 he read the new 6th edition of *An Essay on the Principle of Population*, written in the late 18th century by Thomas Robert Malthus. Malthus' idea of population growth leading to a struggle for survival combined with Darwin's knowledge on how breeders selected traits, led to the inception of Darwin's theory of natural selection. Darwin did not publish his ideas on evolution for 20 years. However, he did share them with certain other naturalists and friends, starting with Joseph Dalton Hooker, with whom he discussed his unpublished 1844 essay on natural selection. During this period he used the time he could spare from his other scientific work to slowly refine his ideas and, aware of the intense controversy around transmutation, amass evidence to support them. In September 1854 he began full-time work on writing his book on natural selection.

Unlike Darwin, Alfred Russel Wallace, influenced by the book *Vestiges of the Natural History of Creation*, already suspected that transmutation of species occurred when he began his career as a naturalist. By 1855, his biogeographical observations during his field work in South America and the Malay Archipelago made him confident enough in a branching pattern of evolution to publish a paper stating that every species originated in close proximity to an already existing closely allied species. Like Darwin, it was Wallace's consideration of how the ideas of Malthus might apply to animal populations that led him to conclusions very similar to those reached by Darwin about the role of natural selection. In February 1858, Wallace, unaware of Darwin's unpublished ideas, composed his thoughts into an essay and mailed them to Darwin, asking for his opinion. The result was the joint publication in July of an extract from Darwin's 1844 essay along with Wallace's letter. Darwin also began work on a short abstract summarising his theory, which he would publish in 1859 as *On the Origin of Species*.

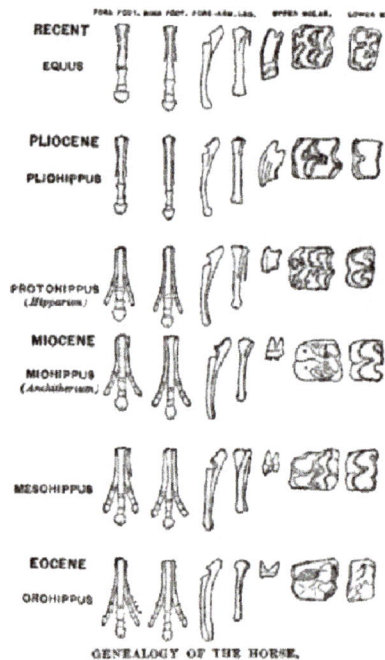

Diagram by Othniel Charles Marsh of the evolution of horse feet and teeth over time as reproduced in Thomas Henry Huxley's *Prof. Huxley in America* (1876)

1859–1930s: Darwin and his Legacy

By the 1850s, whether or not species evolved was a subject of intense debate, with prominent scientists arguing both sides of the issue. The publication of Charles Darwin's *On the Origin of Species* fundamentally transformed the discussion over biological origins. Darwin argued that his branching version of evolution explained a wealth of facts in biogeography, anatomy, embryology, and other fields of biology. He also provided the first cogent mechanism by which evolutionary change could persist: his theory of natural selection.

One of the first and most important naturalists to be convinced by *Origin* of the reality of evolution was the British anatomist Thomas Henry Huxley. Huxley recognized that unlike the earlier transmutational ideas of Jean-Baptiste Lamarck and *Vestiges of the Natural History of Creation*, Darwin's theory provided a mechanism for evolution without supernatural involvement, even if Huxley himself was not completely convinced that natural selection was the key evolutionary mechanism. Huxley would make advocacy of evolution a cornerstone of the program of the X Club to reform and professionalise science by displacing natural theology with naturalism and to end the domination of British natural science by the clergy. By the early 1870s in English-speaking countries, thanks partly to these efforts, evolution had become the mainstream scientific explanation for the origin of species. In his campaign for public and scientific acceptance of Darwin's theory, Huxley made extensive use of new evidence for evolution from paleontology. This included evidence that birds had evolved from reptiles, including the discovery of *Archaeopteryx* in Europe, and a number of fossils of primitive birds with teeth found in North America. Another important line of evidence was the finding of fossils that helped trace the evolution of the horse from its small five-toed ancestors. However, acceptance of evolution among scientists in non-English speaking nations such as France, and the countries of southern Europe and Latin America was slower. An exception to this was Germany, where both August Weismann and Ernst Haeckel championed this idea: Haeckel used evolution to challenge the established tradition of metaphysical idealism in German biology, much as Huxley used it to challenge natural theology in Britain. Haeckel and other German scientists would take the lead in launching an ambitious programme to reconstruct the evolutionary history of life based on morphology and embryology.

Darwin's theory succeeded in profoundly altering scientific opinion regarding the development of life and in producing a small philosophical revolution. However, this theory could not explain several critical components of the evolutionary process. Specifically, Darwin was unable to explain the source of variation in traits within a species, and could not identify a mechanism that could pass traits faithfully from one generation to the next. Darwin's hypothesis of pangenesis, while relying in part on the inheritance of acquired characteristics, proved to be useful for statistical models of evolution that were developed by his cousin Francis Galton and the "biometric" school of evolutionary thought. However, this idea proved to be of little use to other biologists.

Application to Humans

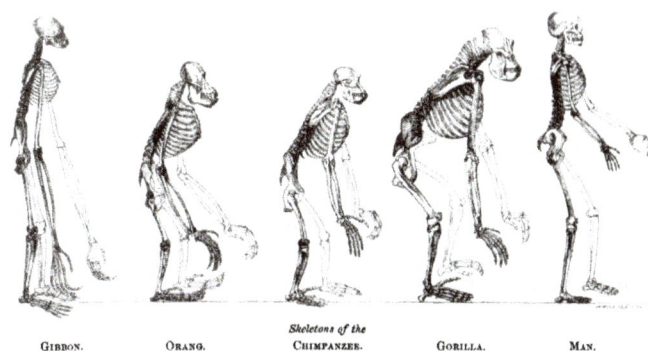

GIBBON.   ORANG.   Skeletons of the CHIMPANZEE.   GORILLA.   MAN.

Photographically reduced from Diagrams of the natural size (except that of the Gibbon, which was twice as large as nature), drawn by Mr. Waterhouse Hawkins from specimens in the Museum of the Royal College of Surgeons.

This illustration was the frontispiece of Thomas Henry Huxley's book *Evidence as to Man's Place in Nature* (1863). Huxley applied Darwin's ideas to humans, using comparative anatomy to show that humans and apes had a common ancestor, which challenged the theologically important idea that humans held a unique place in the universe.

Charles Darwin was aware of the severe reaction in some parts of the scientific community against the suggestion made in *Vestiges of the Natural History of Creation* that humans had arisen from animals by a process of transmutation. Therefore, he almost completely ignored the topic of human evolution in *On the Origin of Species*. Despite this precaution, the issue featured prominently in the debate that followed the book's publication. For most of the first half of the 19th century, the scientific community believed that, although geology had shown that the Earth and life were very old, human beings had appeared suddenly just a few thousand years before the present. However, a series of archaeological discoveries in the 1840s and 1850s showed stone tools associated with the remains of extinct animals. By the early 1860s, as summarized in Charles Lyell's 1863 book *Geological Evidences of the Antiquity of Man*, it had become widely accepted that humans had existed during a prehistoric period—which stretched many thousands of years before the start of written history. This view of human history was more compatible with an evolutionary origin for humanity than was the older view. On the other hand, at that time there was no fossil evidence to demonstrate human evolution. The only human fossils found before the discovery of Java Man in the 1890s were either of anatomically modern humans or of Neanderthals that were too close, especially in the critical characteristic of cranial capacity, to modern humans for them to be convincing intermediates between humans and other primates.

Therefore, the debate that immediately followed the publication of *On the Origin of Species* centered on the similarities and differences between humans and modern apes. Carolus Linnaeus had been criticised in the 18th century for grouping humans and apes together as primates in his ground breaking classification system. Richard Owen vigorously defended the classification suggested by Georges Cuvier and Johann Friedrich Blumenbach that placed humans in a separate order from any of the other mammals, which by the early 19th century had become the orthodox view. On the other hand, Thomas Henry Huxley sought to demonstrate a close anatomical relationship between humans and apes. In one famous incident, which became known as the Great Hippocampus Question, Huxley showed that Owen was mistaken in claiming that the brains of gorillas lacked a structure present in human brains. Huxley summarized his argument in his highly influential 1863 book *Evidence as to Man's Place in Nature*. Another viewpoint was advocated by Lyell and Alfred Russel Wallace. They agreed that humans shared a common ancestor with apes, but questioned whether any purely materialistic mechanism could account for all the differences between humans and apes, especially some aspects of the human mind.

In 1871, Darwin published *The Descent of Man, and Selection in Relation to Sex*, which contained his views on human evolution. Darwin argued that the differences between the human mind and the minds of the higher animals were a matter of degree rather than of kind. For example, he viewed morality as a natural outgrowth of instincts that were beneficial to animals living in social groups. He argued that all the differences between humans and apes were explained by a combination of the selective pressures that came from our ancestors moving from the trees to the plains, and sexual selection. The debate over human origins, and over the degree of human uniqueness continued well into the 20th century.

Alternatives to Natural Selection

The concept of evolution was widely accepted in scientific circles within a few years of the publication of *Origin*, but the acceptance of natural selection as its driving mechanism was much less widespread. The four major alternatives to natural selection in the late 19th century were theistic evolution, neo-Lamarckism, orthogenesis, and saltationism.

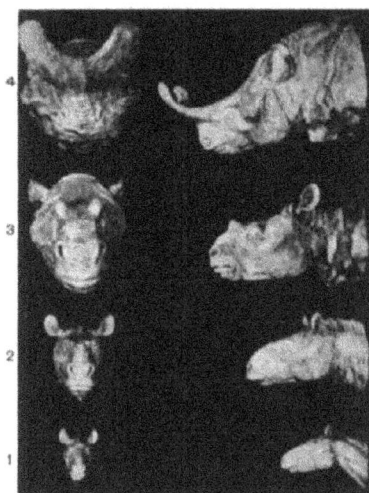

This photo from Henry Fairfield Osborn's 1917 book *Origin and Evolution of Life* shows models depicting the evolution of Titanothere horns over time, which Osborn claimed was an example of an orthogenetic trend in evolution.

Theistic evolution was the idea that God intervened in the process of evolution, to guide it in such a way that the living world could still be considered to be designed. The term was promoted by Charles Darwin's greatest American advocate Asa Gray. However, this idea gradually fell out of favor among scientists, as they became more and more committed to the idea of methodological naturalism and came to believe that direct appeals to supernatural involvement were scientifically unproductive. By 1900, theistic evolution had largely disappeared from professional scientific discussions, although it retained a strong popular following.

In the late 19th century, the term neo-Lamarckism came to be associated with the position of naturalists who viewed the inheritance of acquired characteristics as the most important evolutionary mechanism. Advocates of this position included the British writer and Darwin critic Samuel Butler, the German biologist Ernst Haeckel, and the American paleontologist Edward Drinker Cope. They considered Lamarckism to be philosophically superior to Darwin's idea of selection acting on random variation. Cope looked for, and thought he found, patterns of linear progression in the fossil record. Inheritance of acquired characteristics was part of Haeckel's recapitulation theory of evolution, which held that the embryological development of an organism repeats its evolutionary history. Critics of neo-Lamarckism, such as the German biologist August Weismann and Alfred Russel Wallace, pointed out that no one had ever produced solid evidence for the inheritance of acquired characteristics. Despite these criticisms, neo-Lamarckism remained the most popular alternative to natural selection at the end of the 19th century, and would remain the position of some naturalists well into the 20th century.

Orthogenesis was the hypothesis that life has an innate tendency to change, in a unilinear fashion, towards ever-greater perfection. It had a significant following in the 19th century, and its proponents included the Russian biologist Leo S. Berg and the American paleontologist Henry Fairfield Osborn. Orthogenesis was popular among some paleontologists, who believed that the fossil record showed a gradual and constant unidirectional change.

Saltationism was the idea that new species arise as a result of large mutations. It was seen as a much faster alternative to the Darwinian concept of a gradual process of small random variations being acted on by natural selection, and was popular with early geneticists such as Hugo de Vries, William Bateson, and early in his career, Thomas Hunt Morgan. It became the basis of the mutation theory of evolution.

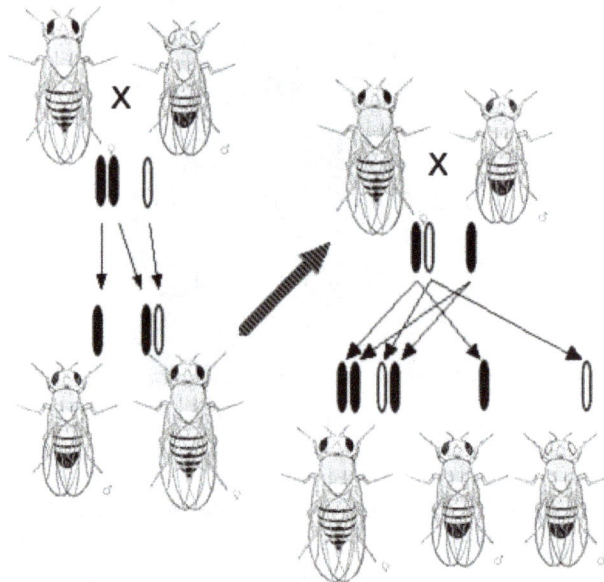

Diagram from Thomas Hunt Morgan's 1919 book *The Physical Basis of Heredity*, showing the sex-linked inheritance of the white-eyed mutation in *Drosophila melanogaster*.

Mendelian Genetics, Biometrics, and Mutation

The rediscovery of Gregor Mendel's laws of inheritance in 1900 ignited a fierce debate between two camps of biologists. In one camp were the Mendelians, who were focused on discrete variations and the laws of inheritance. They were led by William Bateson (who coined the word *genetics*) and Hugo de Vries (who coined the word *mutation*). Their opponents were the biometricians, who were interested in the continuous variation of characteristics within populations. Their leaders, Karl Pearson and Walter Frank Raphael Weldon, followed in the tradition of Francis Galton, who had focused on measurement and statistical analysis of variation within a population. The biometricians rejected Mendelian genetics on the basis that discrete units of heredity, such as genes, could not explain the continuous range of variation seen in real populations. Weldon's work with crabs and snails provided evidence that selection pressure from the environment could shift the range of variation in wild populations, but the Mendelians maintained that the variations measured by biometricians were too insignificant to account for the evolution of new species.

When Thomas Hunt Morgan began experimenting with breeding the fruit fly *Drosophila melanogaster*, he was a saltationist who hoped to demonstrate that a new species could be created in the lab by mutation alone. Instead, the work at his lab between 1910 and 1915 reconfirmed Mendelian genetics and provided solid experimental evidence linking it to chromosomal inheritance. His work also demonstrated that most mutations had relatively small effects, such as a change in eye color, and that rather than creating a new species in a single step, mutations served to increase variation within the existing population.

1920s–1940s

Population Genetics

The Mendelian and biometrician models were eventually reconciled with the development of population genetics. A key step was the work of the British biologist and statistician Ronald Fisher. In a series of papers starting in 1918 and culminating in his 1930 book *The Genetical Theory of Natural Selection*, Fisher showed that the continuous variation measured by the biometricians could be produced by the combined action of many discrete genes, and that natural selection could change gene frequencies in a population, resulting in evolution. In a series of papers beginning in 1924, another British geneticist, J. B. S. Haldane, applied statistical analysis to real-world examples of natural selection, such as the evolution of industrial melanism in peppered moths, and showed that natural selection worked at an even faster rate than Fisher assumed.

The American biologist Sewall Wright, who had a background in animal breeding experiments, focused on combinations of interacting genes, and the effects of inbreeding on small, relatively isolated populations that exhibited genetic drift. In 1932, Wright introduced the concept of an adaptive landscape and argued that genetic drift and inbreeding could drive a small, isolated sub-population away from an adaptive peak, allowing natural selection to drive it towards different adaptive peaks. The work of Fisher, Haldane and Wright founded the discipline of population genetics. This integrated natural selection with Mendelian genetics, which was the critical first step in developing a unified theory of how evolution worked.

Modern Evolutionary Synthesis

In the first few decades of the 20th century, most field naturalists continued to believe that Lamarckian and orthogenetic mechanisms of evolution provided the best explanation for the complexity they observed in the living world. But as the field of genetics continued to develop, those views became less tenable. Theodosius Dobzhansky, a postdoctoral worker in Thomas Hunt Morgan's lab, had been influenced by the work on genetic diversity by Russian geneticists such as Sergei Chetverikov. He helped to bridge the divide between the foundations of microevolution developed by the population geneticists and the patterns of macroevolution observed by field biologists, with his 1937 book *Genetics and the Origin of Species*. Dobzhansky examined the genetic diversity of wild populations and showed that, contrary to the assumptions of the population geneticists, these populations had large amounts of genetic diversity, with marked differences between sub-populations. The book also took the highly mathematical work of the population geneticists and put it into a more accessible form. In Britain, E. B. Ford, the pioneer of ecological genetics, continued throughout the 1930s and 1940s to demonstrate the power of selection due to ecological factors including the ability to maintain genetic diversity through genetic polymorphisms such as human blood types. Ford's work would contribute to a shift in emphasis during the course of the modern synthesis towards natural selection over genetic drift.

Evolutionary biologist Ernst Mayr was influenced by the work of the German biologist Bernhard Rensch showing the influence of local environmental factors on the geographic distribution of sub-species and closely related species. Mayr followed up on Dobzhansky's work with the 1942 book *Systematics and the Origin of Species*, which emphasized the importance of allopatric speciation in the formation of new species. This form of speciation occurs when the geographical isolation of a sub-population is followed by the development of mechanisms for reproductive isolation. Mayr also formulated the biological species concept that defined a species as a group of interbreeding or potentially interbreeding populations that were reproductively isolated from all other populations.

In the 1944 book *Tempo and Mode in Evolution*, George Gaylord Simpson showed that the fossil record was consistent with the irregular non-directional pattern predicted by the developing evolutionary synthesis, and that the linear trends that earlier paleontologists had claimed supported orthogenesis and neo-Lamarckism did not hold up to closer examination. In 1950, G. Ledyard Stebbins published *Variation and Evolution in Plants*, which helped to integrate botany into the synthesis. The emerging cross-disciplinary consensus on the workings of evolution would be known as the modern evolutionary synthesis. It received its name from the 1942 book *Evolution: The Modern Synthesis* by Julian Huxley.

The evolutionary synthesis provided a conceptual core—in particular, natural selection and Mendelian population genetics—that tied together many, but not all, biological disciplines. It helped establish the legitimacy of evolutionary biology, a primarily historical science, in a scientific climate that favored experimental methods over historical ones. The synthesis also resulted in a considerable narrowing of the range of mainstream evolutionary thought (what Stephen Jay Gould called the "hardening of the synthesis"): by the 1950s, natural selection acting on genetic variation was virtually the only acceptable mechanism of evolutionary change (panselectionism), and macroevolution was simply considered the result of extensive microevolution.

1940s–1960s: Molecular Biology and Evolution

The middle decades of the 20th century saw the rise of molecular biology, and with it an understanding of the chemical nature of genes as sequences of DNA and of their relationship—through the genetic code—to protein sequences. At the same time, increasingly powerful techniques for analyzing proteins, such as protein electrophoresis and sequencing, brought biochemical phenomena into realm of the synthetic theory of evolution. In the early 1960s, biochemists Linus Pauling and Emile Zuckerkandl proposed the molecular clock hypothesis (MCH): that sequence differences between homologous proteins could be used to calculate the time since two species diverged. By 1969, Motoo Kimura and others provided a theoretical basis for the molecular clock, arguing that—at the molecular level at least—most genetic mutations are neither harmful nor helpful and that mutation and genetic drift (rather than natural selection) cause a large portion of genetic change: the neutral theory of molecular evolution. Studies of protein differences *within* species also brought molecular data to bear on population genetics by providing estimates of the level of heterozygosity in natural populations.

From the early 1960s, molecular biology was increasingly seen as a threat to the traditional core of evolutionary biology. Established evolutionary biologists—particularly Ernst Mayr, Theodosius Dobzhansky, and George Gaylord Simpson, three of the architects of the modern synthesis—were extremely skeptical of molecular approaches, especially when it came to the connection (or lack thereof) to natural selection. The molecular-clock hypothesis and the neutral theory were particularly controversial, spawning the neutralist-selectionist debate over the relative importance of mutation, drift and selection, which continued into the 1980s without a clear resolution.

Late 20th Century

Gene-Centered View

In the mid-1960s, George C. Williams strongly critiqued explanations of adaptations worded in terms of "survival of the species" (group selection arguments). Such explanations were largely replaced by a gene-centered view of evolution, epitomized by the kin selection arguments of W. D. Hamilton, George R. Price and John Maynard Smith. This viewpoint would be summarized and popularized in the influential 1976 book *The Selfish Gene* by Richard Dawkins. Models of the period seemed to show that group selection was severely limited in its strength; though newer models do admit the possibility of significant multi-level selection.

In 1973, Leigh Van Valen proposed the term "Red Queen," which he took from *Through the Looking-Glass* by Lewis Carroll, to describe a scenario where a species involved in one or more evolutionary arms races would have to constantly change just to keep pace with the species with which it was co-evolving. Hamilton, Williams and others suggested that this idea might explain the evolution of sexual reproduction: the increased genetic diversity caused by sexual reproduction would help maintain resistance against rapidly evolving parasites, thus making sexual reproduction common, despite the tremendous cost from the gene-centric point of view of a system where only half of an organism's genome is passed on during reproduction.

However, contrary to the expectations of the Red Queen hypothesis, Hanley *et al.* found that the prevalence, abundance and mean intensity of mites was significantly higher in sexual geckos than in asexuals sharing the same habitat. Furthermore, Parker, after reviewing numerous genetic studies on plant disease resistance, failed to find a single example consistent with the concept that pathogens are the primary selective agent responsible for sexual reproduction in their host. At an even more fundamental level, Heng and Gorelick and Heng reviewed evidence that sex, rather than enhancing diversity, acts as a constraint on genetic diversity. They considered that sex acts as a coarse filter, weeding out major genetic changes, such as chromosomal rearrangements, but permitting minor variation, such as changes at the nucleotide or gene level (that are often neutral) to pass through the sexual sieve. The adaptive function of sex, today, remains a major unresolved issue in biology. The competing models to explain the adaptive function of sex were reviewed by Birdsell and Wills. A principal alternative view to the Red Queen hypothesis is that sex arose, and is maintained, as a process for repairing DNA damage, and that genetic variation is produced as a byproduct.

The gene-centric view has also led to an increased interest in Charles Darwin's old idea of sexual selection, and more recently in topics such as sexual conflict and intragenomic conflict.

Sociobiology

W. D. Hamilton's work on kin selection contributed to the emergence of the discipline of sociobiology. The existence of altruistic behaviors has been a difficult problem for evolutionary theorists from the beginning. Significant progress was made in 1964 when Hamilton formulated the inequality in kin selection known as Hamilton's rule, which showed how eusociality in insects (the existence of sterile worker classes) and many other examples of altruistic behavior could have evolved through kin selection. Other theories followed, some derived from game theory, such as reciprocal altruism. In 1975, E. O. Wilson published the influential and highly controversial book *Sociobiology: The New Synthesis* which claimed evolutionary theory could help explain many aspects of animal, including human, behavior. Critics of sociobiology, including Stephen Jay Gould and Richard Lewontin, claimed that sociobiology greatly overstated the degree to which complex human behaviors could be determined by genetic factors. They also claimed that the theories of sociobiologists often reflected their own ideological biases. Despite these criticisms, work has continued in sociobiology and the related discipline of evolutionary psychology, including work on other aspects of the altruism problem.

Evolutionary Paths and Processes

One of the most prominent debates arising during the 1970s was over the theory of punctuated equilibrium. Niles Eldredge and Stephen Jay Gould proposed that there was a pattern of fossil species that remained largely unchanged for long periods (what they termed *stasis*), interspersed with relatively brief periods of rapid change during speciation. Improvements in sequencing methods resulted in a large increase of sequenced genomes, allowing the testing and refining of evolutionary theories using this huge amount of genome data. Comparisons between these genomes provide insights into the molecular mechanisms of speciation and adaptation. These genomic analyses have produced fundamental changes in the understanding of the evolutionary history of life, such as the proposal of the three-domain system by Carl Woese. Advances in computational hardware and software allow the testing and extrapolation of increasingly advanced evolutionary models and the development of the field of systems biology. One of the results has been an exchange of ideas between theories of biological evolution and the field of computer science known as evolutionary computation, which attempts to mimic biological evolution for the purpose of developing new computer algorithms. Discoveries in biotechnology now allow the modification of entire genomes, advancing evolutionary studies to the level where future experiments may involve the creation of entirely synthetic organisms.

Microbiology, Horizontal Gene Transfer, and Endosymbiosis

Microbiology was largely ignored by early evolutionary theory. This was due to the paucity of morphological traits and the lack of a species concept in microbiology, particularly amongst prokaryotes. Now, evolutionary researchers are taking advantage of their improved understanding of microbial physiology and ecology, produced by the comparative ease of microbial genomics, to explore the taxonomy and evolution of these organisms. These studies are revealing unanticipated levels of diversity amongst microbes.

One important development in the study of microbial evolution came with the discovery in Japan in 1959 of horizontal gene transfer. This transfer of genetic material between different species of bacteria came to the attention of scientists because it played a major role in the spread of antibiotic resistance. More recently, as knowledge of genomes has continued to expand, it has been suggested that lateral transfer of genetic material has played an important role in the evolution of all organisms. These high levels of horizontal gene transfer have led to suggestions that the family tree of today's organisms, the so-called "tree of life," is more similar to an interconnected web or net.

Indeed, the endosymbiotic theory for the origin of organelles sees a form of horizontal gene transfer as a critical step in the evolution of eukaryotes such as fungi, plants, and animals. The endosymbiotic theory holds that organelles within the cells of eukorytes such as mitochondria and chloroplasts, had descended from independent bacteria that came to live symbiotically within other cells. It had been suggested in the late 19th century when similarities between mitochondria and bacteria were noted, but largely dismissed until it was revived and championed by Lynn Margulis in the 1960s and 1970s; Margulis was able to make use of new evidence that such organelles had their own DNA that was inherited independently from that in the cell's nucleus.

Evolutionary Developmental Biology

In the 1980s and 1990s, the tenets of the modern evolutionary synthesis came under increasing scrutiny. There was a renewal of structuralist themes in evolutionary biology in the work of biologists such as Brian Goodwin and Stuart Kauffman, which incorporated ideas from cybernetics and systems theory, and emphasized the self-organizing processes of development as factors directing the course of evolution. The evolutionary biologist Stephen Jay Gould revived earlier ideas of heterochrony, alterations in the relative rates of developmental processes over the course of evolution, to account for the generation of novel forms, and, with the evolutionary biologist Richard Lewontin, wrote an influential paper in 1979 suggesting that a change in one biological structure, or even a structural novelty, could arise incidentally as an accidental result of selection on another structure, rather than through direct selection for that particular adaptation. They called such incidental structural changes "spandrels" after an architectural feature. Later, Gould and Elisabeth Vrba discussed the acquisition of new functions by novel structures arising in this fashion, calling them "exaptations."

Molecular data regarding the mechanisms underlying development accumulated rapidly during the 1980s and 1990s. It became clear that the diversity of animal morphology was not the result of different sets of proteins regulating the development of different animals, but from changes in the deployment of a small set of proteins that were common to all animals. These proteins became known as the "developmental-genetic toolkit." Such perspectives influenced the disciplines of phylogenetics, paleontology and comparative developmental biology, and spawned the new discipline of evolutionary developmental biology also known as evo-devo.

21st Century

Macroevolution and Microevolution

One of the tenets of the modern evolutionary synthesis was that macroevolution (the evolution of phylogenic clades at the species level and above) was solely the result of the mechanisms of microevolution (changes in gene frequency within populations) operating over an extended period of time. During the last decades of the 20th century some paleontologists raised questions about whether other factors, such as punctuated equilibrium and group selection operating on the level of entire species and even higher level phylogenic clades, needed to be considered to explain patterns in evolution revealed by statistical analysis of the fossil record. Near the end of the 20th century some researchers in evolutionary developmental biology suggested that interactions between the environment and the developmental process might have been the source of some of the structural innovations seen in macroevolution, but other evo-devo researchers maintained that genetic mechanisms visible at the population level are fully sufficient to explain all macroevolution.

Epigenetic Inheritance

Epigenetics is the study of heritable changes in gene expression or cellular phenotype caused by mechanisms other than changes in the underlying DNA sequence. By the first decade of the 21st century it had become accepted that epigenetic mechanisms were a necessary part of the evolutionary origin of cellular differentiation. Although epigenetics in multicellular organisms is generally thought to be a mechanism involved in differentiation, with epigenetic patterns "reset" when organisms reproduce, there have been some observations of transgenerational epigenetic inheritance. This shows that in some cases nongenetic changes to an organism can be inherited and it has been suggested that such inheritance can help with adaptation to local conditions and affect evolution. Some have suggested that in certain cases a form of Lamarckian evolution may occur.

Unconventional Evolutionary Theory

Omega Point

Pierre Teilhard de Chardin's metaphysical Omega Point theory, found in his book *The Phenomenon of Man* (1955), describes the gradual development of the universe from subatomic particles to human society, which he viewed as its final stage and goal.

Gaia Hypothesis

Teilhard de Chardin's ideas have been seen by advocates of the Gaia hypothesis proposed by James Lovelock, which holds that the living and nonliving parts of Earth can be viewed as a complex interacting system with similarities to a single organism, as being connected to Lovelock's ideas. The Gaia hypothesis has also been viewed by Lynn Margulis and others as an extension of endosymbiosis and exosymbiosis. This modified hypothesis postulates that all living things have a regulatory effect on the Earth's environment that promotes life overall.

Extended Evolutionary Synthesis

The extended evolutionary synthesis (EES) is an extension of the Modern Synthesis of evolution which revisits the relative importance of different factors at play in evolutionary theory. EES includes concepts and mechanisms such as multilevel selection theory, transgenerational epigenetic inheritance, niche construction and evolvability.

References

- Moran, Laurence A. (2006). "Random Genetic Drift". What is Evolution?. Toronto, Canada: University Toronto. Retrieved 2015-09-27.

- Teilhard de Chardin 1959 Abbatucci, Jacques Severin. "Teilhard de Chardin: The Phenomenon of Man: a Compendium". Retrieved 2015-06-15.

- Fox, Robin (December 2004). "Symbiogenesis". Journal of the Royal Society of Medicine. London: Sage Publications. 97 (12): 559. doi:10.1258/jrsm.97.12.559. ISSN 0141-0768. PMC 1079665. PMID 15574850. Retrieved 2014-11-05.

- Miller, James. "Daoism and Nature" (PDF). Archived from the original (PDF) on 2008-12-16. Retrieved 2014-10-26. "Notes for a lecture delivered to the Royal Asiatic Society, Shanghai on January 8, 2008"

- Sedley, David (August 10, 2013). "Lucretius". In Zalta, Edward N. Stanford Encyclopedia of Philosophy (Fall 2013 ed.). Stanford, CA: Stanford University. ISSN 1095-5054. Retrieved 2014-10-26.

- Simpson, David (2006). "Lucretius". Internet Encyclopedia of Philosophy. Martin, TN: University of Tennessee at Martin. ISSN 2161-0002. OCLC 37741658. Retrieved 2014-10-26.

- Carroll, William E. (2000). "Creation, Evolution, and Thomas Aquinas". Revue des Questions Scientifiques. Namur, Belgium: Scientific Society of Brussels. 171 (4). ISSN 0035-2160. Retrieved 2014-10-28.

- Gould, Stephen Jay; Vrba, Elisabeth S. (Winter 1982). "Exaptation—a missing term in the science of form" (PDF). Paleobiology. Paleontological Society. 8 (1): 4–15. ISSN 0094-8373. JSTOR 2400563. Retrieved 2014-11-04.

- Singer, Emily (February 4, 2009). "A Comeback for Lamarckian Evolution?". technologyreview.com. Cambridge, MA: Technology Review, Inc. ISSN 0040-1692. Retrieved 2014-11-05.

- Irvine, Chris (February 11, 2009). "The Vatican claims Darwin's theory of evolution is compatible with Christianity". The Daily Telegraph. London: Telegraph Media Group. Retrieved 2014-10-26.

- Wilkins, John (July–August 2006). "Species, Kinds, and Evolution". Reports of the National Center for Science Education. Berkeley, CA: National Center for Science Education. 26 (4): 36–45. ISSN 2158-818X. Retrieved 2011-09-23.

- Boylan, Michael (September 26, 2005). "Aristotle: Biology". Internet Encyclopedia of Philosophy. Martin, TN: University of Tennessee at Martin. ISSN 2161-0002. OCLC 37741658. Retrieved 2011-09-25.

- Waggoner, Ben. "Medieval and Renaissance Concepts of Evolution and Paleontology". University of California Museum of Paleontology. Retrieved 2010-03-11.

Permissions

Index